SPA PA…
BRI…

Parks and Planting for 'Taking the Waters'

KAREN EMERY

2025

First published in Great Britain 2025

ISBN: 978-1-0369-1223-9

Printed in Great Britain by
Book Printing UK
www.bookprintinguk.com
Remus House, Coltsfoot Drive, Peterborough, PE2 9BF

In introducing, therefore, for the first time, into the literature of this country a performance of this nature, I must trust to the candour of my readers, that, for the sake of its good intention, they will forgive any defect that may be found in its plan or execution.

Augustus Bozzi Granville, M.D., F.R.S.
The Spas of England and Principal Seabathing Places, 1841

For Hilke

Contents

Preface

Traditional spas – those places where dubious-tasting waters were drunk as part of an extended health cure away from home – are no longer part of our way of life, having been superseded by 'pampering' experiences in luxurious surroundings. Yet former spa culture has left its mark in many places in Britain and Europe. Growing up in Germany, the author has experienced the culture of 'taking the waters' at first hand and has fond early memories of the spa park (*Kurpark)* in the seabathing spa of Norderney (an island off the north coast of Germany).

The rituals around using spa waters invariably involve the provision of supporting facilities. There is, for example, a body of literature on architecture for spa buildings and we see former pump rooms and baths as remaining centrepieces in spa towns like Bath and Buxton. The attractive surroundings, landscaping and leafy avenues are another feature characteristic of many spa towns. A strong tradition of health tourism at spa resorts in continental Europe has actually identified the '*Kurpark*' and *'Kurgarten'* (spa park and spa garden) as a subtype in garden history, in which the planted environment is an integral part of the visitor experience that was part of 'the cure'.[1] Here in Britain, the development of such spa landscape has not been given the same attention and is confined to description of individual places without an overview that may link common themes. This book is intended to address this lack and to compile known information on parks, gardens and landscape features that were designed or adapted specifically for a spa-visiting public and to raise awareness of their contribution to the development of public parks.

PART I

The Study of Spa Landscapes

Garden history should concern itself with those aspects of the landscape over which humans have had some control to shape and order the land. [2] Whilst this seems fairly straightforward, the interpretation of the history of parks and gardens has gone through a number of phases, the approach changing over time. Roy Strong, initiator of the 1979 Victoria and Albert Museum's exhibition *The Garden: A Celebration of One Thousand Years of British Gardening*, noted at the time that 'as an area of academic study, garden history is a relatively new one'.[3]

The study of gardens has traditionally focused on aesthetic design, looking at the development of structure and planting schemes as artistic endeavour rather than how those spaces were used for specific purposes. But just like the study of history generally, which moved from mere description of people and dates to considering the context and the 'whys and wherefores', garden historians have begun to consider the sociological aspects of garden design. A more modern approach is to recognise that gardens and parks are an integral component of the culture and society in which they are made, reflecting the lives and aspirations of people at various points in time. There are the private parks and gardens that are created for personal pleasure, self-expression or confirmation of personal status. There are designed landscapes that reflect beliefs or affiliations and broadcast them to others. And then there are functional spaces, planted to incorporate aesthetically pleasing elements as well as serving a purpose. Spa parks definitely fall into this last category.

In terms of public green spaces, it has been identified that the focus has been limited so far to the most obvious and widely studied categories: eighteenth century pleasure gardens and Victorian public parks.

> *The two periods and urban types that have come to dominate studies of historical urban greenspace are 'The Georgian Town' and 'The Victorian City'. Consequently, recognition of the significance of historic public greenspace tends to be confined to pleasure gardens and municipal parks.'*[4]

This quote from a 2014 English Heritage Research Report confirms that a wider range of purpose and intent for public green spaces has yet to be recognised in garden history, one of which is the development of landscapes designed for spas.

When it comes to planting for public consumption, the intention and justification of the provision is a key factor and this intersects with the commercial aspects of garden design. As will be discussed later, the function of spa surroundings was to provide a setting for exercise and relaxation. But also, the culture of spa visiting involved a section of society that had an interest in refined living. Those larger spas with royal patronage would expect to be able to offer an environment acceptable to the sensibilities of visitors used to sophisticated architecture, fine dining, musical entertainment and, yes: attractive outdoor spaces in which to promenade and spend one's time whilst taking 'the cure'. Minor spas often aspired to advance their reputation through development of spa parks - an overt demonstration of their ambition to provide suitably elegant facilities. Therefore, spa landscapes are associated with what we might term transactional motives, the direct or indirect financial benefit from creating these parks and gardens.

Unfortunately 'greenery' is particularly prone to decay or alteration over time, so evidence of past gardens is rather more difficult to find and interpret than, say, buildings. We rely on individuals who happened to include description of landscaping in their writings, drawings or photographs to get a picture of what these park or garden features looked like and what impact they had on their audience. Irritatingly - for garden historians - it appears that outdoor spaces were taken for granted, often not worthy of specific mention. It is, for example, astonishing how often the seventeenth century diarist Samuel Pepys mentions that he 'walked in the garden' with various personages, without telling us anything about that garden!

People who visited spas for treatments mostly had their own ailments and the attractions of the social scene in the forefront of their minds. This means that those who wrote about their spa experiences generally focused on personalities, the built environment and the wider scenery, leaving fewer descriptions of the immediate outdoor environment in which they walked, talked, schemed and sometimes duelled. It appears that the parks and avenues were just a normal background that contributed to the general ambience of the spa. Nevertheless, the existence of scenic and leafy environs was, at least subconsciously, an important element of the spa experience.

The German garden writer Christian Hirschfeld (1742 – 1792) was one of very few who wrote of spa parks and gardens as a specific type. He maintained that consideration should be given to provide a suitable environment for the health-seeker. As an enthusiast for what came to be known as the 'English Landscape Garden', he was keen on vistas and the contribution of the wider, unencumbered view of the scenic surroundings. A fascinating choice in his book for an English spa is Matlock Bath in Derbyshire. This spa developed during the late seventeenth and from the beginning, the dramatic scenery was as much an attraction as the waters. Over more than two pages of his 1785 publication, Hirschfeld expressed his delight over the natural scenic beauty of the environment of the Derwent Valley. He did however bemoan that very little effort had been expended to make more of this landscape. He wished *'dass die Kunst etwas dazu beitragen möchte, um sie in ihrer ganzen Schönheit zu zeigen'* (that art would be allowed to contribute to display it in its full beauty). It may well be that what he experienced was not of a high enough standard for his ideas of beautification, but with so many other spas to choose from, perhaps he did not make a sufficiently thorough survey of British spas to do them justice.

Nineteenth century etching of the scenic Derwent Valley over Matlock.

The academic exploration of the leisure and tourism in Britain has touched on the topic of spa landscape. It has been noted that in the early stages of their development, watering places were locations away from the town and their rural aspects were highlighted as an antidote to urban living. In his *History of Leisure* Peter Borsay notes that the spas

> *...cultivated a pastoral strand in their make-up, with walks, pleasure gardens, parks and forms of residential architecture which embraced the natural environs.*[5]

The modern impetus for reviewing the specific character of spa landscapes also comes from the European continent: Fred Kaspar, a German art historian has, since the 1990s, made extensive studies of spa culture in northwest Europe. His fascination with Westphalian provincial architecture and spas led him to provide an overview of

Kurgärten (spa gardens), which he characterised as an initial 'sketch' since 'the history of parks and green spaces in spa towns and resorts has not been written as yet'.[6] His short bilingual publication of 2016 outlined some common themes in the development of spas and their planned landscapes. A spur for collating research on spa landscapes was a symposium held at the German spa Bad Kissingen in 2014. It was part of the serial transnational nomination for inscription on the UNESCO (United Nations Educational, Scientific and Cultural Organisation) World Heritage list as 'Great Spas of Europe' (ratified in July 2021) which includes Bath spa in England as one of eleven European spas.[7] This was followed up in 2015 by the ICOMOS (International Council on Monuments and Sites) conference, which produced an associated publication in which the UK's contribution was a presentation on Bath[8] and a paper on the role of music, a related strand of spa life that was often offered outdoors in the park.[9]

Other published work on spa gardens is scarce. In Britain, the academic journal *Garden History* has included just one contribution relating to spa landscape, which featured Ronneby Spa in Sweden.[10] In Spain, a 2019 symposium on spa 'thermalism' included a paper on parks and gardens in spa towns.[11] Many of the papers on individual European spas come to conclusions that relate to their own national traditions of spa therapy and urban development, not all of which coincide with the British experience. Although such case studies allow an exchange of information, it was noted that monographic approaches are still predominant, discussing individual examples rather than attempting to extract common principles for the topic.[12]

The study of the 'Vauxhall'-type pleasure gardens, designed for the entertainment of a visiting public from the mid-seventeenth century onwards, is one that has been extensively explored, no doubt due to the fascination of the grand outdoor spectacles and entertainment which left copious reports for historians to mull over. This research can overlap with the study of spa gardens for the social aspects of life in spa towns and indeed the early seaside resorts.[13]

The culture of spas, pleasure gardens and teagardens around the metropolis of London has been thoroughly documented, noting that the healing aspect of the waters was often a minor consideration for their visitors.[14] The history of garden-based entertainment including orchestras and bandstands has also been considered in a number of publications, including Paul Rabbitt's 2018 book on bandstands in Britain. His book is one of very few acknowledging spa parks as such and identifying that the provision of music in spa parks had a role in shaping 'park life'.[15]

Whilst the pleasure gardens of the eighteenth century were entirely commercial fun-parks for the elite, the kind of municipal parks we are all used to, with paths and ponds, fountains and flowers, bandstands and sports facilities were ostensibly designed for health, well-being and raising standards of behaviour for the growing population in Victorian times. Emphasis was on space and facilities for 'appropriate' interactions of local residents. The worthy but certainly paternalistic idea was to engage workers in 'rational recreation' rather than the bawdy behaviour created by that other leisure-time occupation: ale house drinking. This idea of 'Peoples Parks' is a topic comprehensively presented in the recent update of Hazel Conway's original 1980s book on Victorian public parks by Paul Rabbitts and others.[16]

It is often considered that the first municipal parks date from the early 1840s, with Derby Arboretum (1840), Princes Park, Liverpool (1842) and Birkenhead Park (1847) in the vanguard. Yet the twentieth century garden historian Christopher Thacker considered parks in the spa towns of Tunbridge Wells, Cheltenham and Bath as among the earliest examples of public parks.[17] This is an interesting observation, as they were all created *before* the famous 1833 speech by Robert Slaney, MP, to the parliamentary Public Health Committee.[18] His presentation had suggested the securing of open spaces as public walks, a landmark event that is deemed to have been the impetus for the creation of public parks as we know them. The fact that Thacker's three mentioned early parks are – or at least were – located in the top spa resorts in Britain is indicative of their distinctive origin, designed for health visitors rather than the local population. They are true examples of the *Kurparks, i.e.* the spa parks. Whilst Thacker did not by any means claim his list of early parks was exhaustive – it was meant to demonstrate the variety of parks in existence – there are other spa-based examples of what we might regard as 'pioneer parks', established before the public park revolution.

In Britain, many of the earliest spa parks were so obliterated by building or transport development that no vestiges of planting remain. Others morphed into municipal parks, where the spa park characteristics merged into features that typified the nineteenth century civic park, or they reverted to private gardens. As the label of 'spa park' has entirely to do with the culture prevailing at the time that it was created, it is inevitable that once a location ceased to function as a spa, its park was, strictly speaking, no longer a spa park. This may well explain why in Britain spa landscaping has generally not been considered separately from the study of public parks.

Gloucester Public Park, the former 'Rigney Stile Spa Pleasure Grounds' (p.78).

Is it valid to consider spa landscaping separately from other parks and avenues? After all, the arboriculturist Mark Johnston, in writing about tree planting in urban settings, dismisses spa parks as a category because 'these landscapes were not significantly different from other fashionable towns that were not spa resorts'.[19] But although landscape enhancements had been made in a number of towns to raise their fashionable status, it was in spa towns in particular that there was the impetus to provide 'public walks'. A stay of several weeks in a 'watering place' became a regular feature of life for those who could afford it and spas great and small grew up providing facilities for this form of tourism. As for comparison with public parks, the German writer of *Spa Gardens*, Fred Kaspar, cautions against considering spa parks in connection with municipal parks, asserting that 'both types of public green space serve similar but nevertheless functionally different purposes'.[20] These differences in intent and function are summarised on pages 15 and 16.

In truth, the distinct nature of spas and their *Kurparks* with their underlying health objective was more pronounced - and longer-lived - on the mainland of Europe than in Britain Yet the development of spa parks prior to the public park movement demonstrates that spa culture provided a particular pattern for public leisure that is worthy of recognition.

A spa park in full use with promenading, Bath chair and musical entertainment.

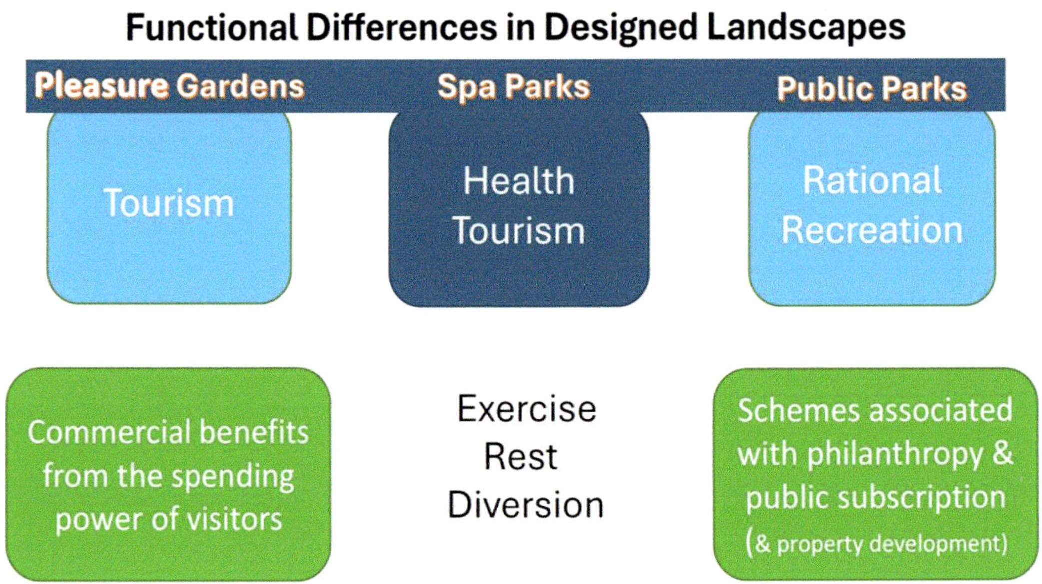
Functional Differences in Designed Landscapes
Pleasure Gardens
Spa Parks
Public Parks
Tourism
Health Tourism
Rational Recreation
Commercial benefits from the spending power of visitors
Exercise
Rest
Diversion
Schemes associated with philanthropy & public subscription
(& property development)

The function and operation of
Spa Parks versus Municipal Public Parks.

Function	***Spa Park***	***Municipal Public Park***
Health	Outdoor space to support spa-based requirement for exercise and rest as essential element of a 'cure' for health visitors.	Outdoor breathing space and physical recreation opportunity for urban residents to enhance public health.
Social	Environment and facilities conducive to pleasant diversion for health tourists.	Space and facilities for appropriate interactions as rational recreation for the resident population.
Commercial	Events, activities and performances as commercial enterprises.	Opportunities for performances and events that could raise funds to justify upkeep of the park.
Reputation	A backdrop to enhance the desirability of the location as a destination for visitors.	A public emblem of credibility of a modern, successful town or city.
Funding rationalisation	Commercial benefits from the spending power of health visitors to the spa.	Schemes associated with residential property, philanthropy and public subscription.

Spas in Great Britain

Mineral springs arise in regions where there are aquifers - water storing rocks - where it is possible for water to be held within the earth and take up mineral constituents from the rocks as it travels through. In regions of naturally higher ground temperature, the water may emerge as thermal springs, as at Bath, in the west of England where the temperature of Bath spa water averages 45°C. Often the waters are rich in salt – sodium chloride – but other minerals such as magnesium and calcium are also dissolved in these waters. Prized and of particular interest were the iron-rich springs, termed 'chalybeate' from a Latin word Chalybs meaning steel and deriving from the Greek from Chalybes, an ancient people in Asia Minor. The red iron staining around these springs gave additional *frisson* to these waters, in medieval Canterbury giving rise to the belief that the martyr Thomas Becket's blood imbued this 'holy water.' Another constituent considered valuable was the foul-smelling sulphurated water. At Askern near Doncaster a sulphur-rich spring even spawned the epigram:

The devil when passing by Askeron
Was ask-ed what he thought thereon:
Quoth Satan – "Judging from the stink,
I can't be far from home, I think.[21]

Unsurprisingly, this kind of water was considered particularly useful for 'purging'. Unlikely as it may seem, it is possible to have springs with different constituents within reasonable proximity, as the attitude of rock layers is seldom even in any location and groundwater can take different pathways through the rocks. This known fact often led enterprising spa establishments to search for more springs to extend their offer or to go in competition with another nearby mineral well.

Early health tourism started as travel which might be likened to medieval pilgrimage. The aim was to effect healing through waters imbued with special properties at the location they emanate from the ground, originally considered to be 'Holy Wells'. One of countless examples is at Thorpe Arch, near Boston Spa, which had St Helen's Well near an ancient chapel and cross that is said to have been a respected holy site, venerated long before the Romans arrived there.[22]

The insistence that the efficacy of the waters depended on the location started as a belief in the spirituality of the site. This almost instinctive acceptance of the importance of 'place' was perpetuated by physicians who claimed that the healing constituents would dissipate within a short time. They emphasised the importance of the water's original temperature and dissolved gases, which would fade away if the water were transported. That inevitably meant travel and sojourn in a setting away from home – an important precursor to the kind of tourism we have all become used to. The attraction of a visit to a new location is what has been described in studies of leisure activities as exploration of 'the other' space – both physically and in the sense of away from the social norms one is used to a home.[23] This idea was certainly exploited by those spas that had natural scenery to complement their offer of mineral springs and visitor facilities. Some travellers went to considerable lengths to reach far out-of-the-way places such as Llangammarch Wells in mid-Wales or indeed Strathpeffer in the Scottish Highlands.

Moving away from the idea of holy wells, we can define spas as places where the waters were *believed* to have medicinal rather than miraculous properties. A description of the effects of its waters include:

> *...that it is generally understood, in the gout they strengthen the habit; in rheumatisms of long standing, the Bath waters afford most effectual relief; and in indigestion, they powerfully alleviate the distressing sinkings of a disordered stomach. In some cases of the palsy, they also tend to give new power of motion to the affected limbs; and, in the obstruction and irregularities of the female habit, the internal use of these waters seldom fail in removing that pallid and sickly appearance which these disorders generally occasion...*[24]

Copious pages listing chemical constituents of different waters and their supposed effects attest to a belief in cures with a scientific basis. The physical effects of iron-rich, saline or sulphurous waters experienced by patients were described as evidence of efficacy through promotion of 'critical discharge from the emunctories[i] of the body'. For who has not been healed,

> *...without either perspiration, expectoration, or salivation; increased action of the intestines or of the kidneys; suppuration[ii], boils, or eruptions; deposition of gouty matter, or discharge of hemorrhoidal blood?*[25]

However, compiling a comprehensive list of those places in Britain that have at any time been considered spa resorts is not as straightforward as it appears. From the time of the Restoration of the monarchy of Charles II in 1760 to the Regency period of George IV in the first quarter of the eighteen hundreds, around 130 new spas were founded and more than three hundred independent treatises were published on various waters.[26] Early nineteenth century books on spas typically presented detailed – if selective - information on spas and an analysis of the mineral constituents of their waters.[27] We turn in particular to a physician by the name of Dr Augustus Bozzi Granville (hereafter referred to as A. B. Granville) who in the first years of the 1840s wrote a series of books on English spas based on his extensive travels, personally visiting as he reports 'between seventy and eighty places'. His first book on this subject, *Spas of Germany* published in 1837, appears to have been very well received, being written in a conversational style as a kind of travelogue incorporating information on the chemistry of the water, facilities at the spas, cost of lodgings and many other pronouncements on the quality of persons or diversions available. It is an entertaining read of an opinionated man whose guides to spas in both Europe and Britain provide an important window on the spa business as it existed right at the start of the Victorian era.

---

[i] Waste-dispelling organs

[ii] Discharge of pus

Originally, he spent pages ruminating upon the new mode of transport - rail travel – which he initially regarded with caution for use by invalids but came to appreciate it by the time he published his last volume on 'Southern Spas'. For the purposes of this book, his three volumes of *The Spas of England and Principal Sea Bathing Places,* published in 1841, were invaluable in identifying the spas that he considered to be active at that time, with copious details relating to both the physical and social landscape of the time.

Twentieth century authors on spas, notably Phyllis Hembry, have generally considered around 130 different former spa locations.[28] However, maps and local histories often show up individual wells that are not part of such compilations and a 2022 study of British spas in eighteenth century medicine and literature has identified a total of 346.[29]

In the search for spas in Britain, placenames can indicate the former existence of a spa or healing water, such as *Spa* Lane, Stockwell or *Well* ingborough. But some 'spa' wells were in fact no more than springs that would lend extra interest to a commercial teagarden to publicise the establishment.[30] Sometimes a discovered well was utilised only very locally or the intended development of a discovered well did not succeed in drawing visitors for more than a few years. In fact, to be considered as a spa for the purposes of this study, it should really have evidence of substantial numbers of visitors travelling there specifically for 'taking the waters' or for the secondary 'social and recreational' reason of following a fashion for spa visiting. However, to illustrate the range of projects, a few more transitory spas have been included, such as Winkfield Spa near Windsor (page 106). In general, the focus is on the recognised national spa towns such as Buxton, Harrogate or Cheltenham, plus numerous regional centres which became well known for a while and often had royal or noble patronage.

There are seaside spa wells at various locations (including for example Scarborough, Spittal and Hove) and their spa-related landscapes have indeed been considered in this publication. However, the issue of coastal spa resorts is a little complicated, as sea bathing (and sea water drinking) for medicinal reasons in the eighteenth century morphed into outings and holidays to the seaside for leisure purposes only. Whilst drinking and bathing in seawater, first recommended by Dr. Richard Russell in 1753 was as valid a health cure as mineral water (for example, Margate's Sea Bathing Hospital from 1791), the expansion of railway travel from the 1840s encouraged greater emphasis on the leisure element. So, from Victorian times and into the twentieth century, coastal parks and gardens became adornment for sea-side resorts without the associated spa functions. Here Britain differs from Europe, as the continental coastal resorts continued to be regarded as places for 'healthful retreat' right into the twentieth century and therefore retained features typical of spas generally. The *Kurgarten* complete with *Kursaal* (assembly and reading rooms), covered walkways and bandstand performance area on the island resorts of Friesland in Germany still served direct spa-related functions into the 1970s. By contrast, the 1901 '*Kursaal* (named using the German word) at London's favourite seaside destination of Southend in Essex became part of the largest amusement park in southern England, with no facilities for health related activities whatsoever.

Vintage postcard, South Bay Scarborough: Spa and landscaped walks.

Whilst eighteenth-century seaside watering places with their emphasis on the bathing ritual for health could be considered a sub-category of spa towns, here we will confine ourselves to spas with *fresh water* mineral springs only.[31] This would, for example, include Scarborough although even here the planted areas were arguably supporting the sea bathing industry more so than the mineral waters spa.

The vogue for spa-going by people who could afford the time and wealth required for travel and often extended stay brought development opportunities for places blessed with a mineral well. For towns, the building and town planning involved in the development of a successful spa has been seen as a contribution to developing urbanism and architectural style. The city of Bath, 'The Queen of Spas,' famously enhanced its standing through eighteenth century town planning that gave vistas to the surrounding hills from terraces for promenading spa visitors. Spa buildings such as pump rooms, assembly rooms, colonnades, and their associated avenues formed an assemblage that epitomised spa towns and led to expansion into grand architectural lodgings and town houses in leafy quarters of town. The spa townscape invariably included avenues, plant embellishments, gardens and walks. It has been noted that these towns therefore 'played to a green agenda,' providing urban green space which might not otherwise have been incorporated.

It was not uncommon to find spa wells within private gardens as a feature in one's private estate and the existence of a mineral spring was considered a significant attraction for potential buyers. In West Kent in 1841 the Birchden and Hambell Estate scenery was described in auction particulars in glowing terms including 'a mineral spring equal to that at Tunbridge-wells' (sic).[32] Not ten miles away, a chalybeate spring was incorporated into Dunorlan Park, a fine private park landscaped for Henry Reed in the 1860s by Robert Marnock, one of the leading landscape designers of his day. Nowadays this park is a delightful public park with a spa well, rather than a spa with a landscaped park. Even into the twentieth century the Thames-side spring at Cliveden House in Buckinghamshire became a private spa retreat for successive owners from at least the eighteenth century.

Owners of commercial nursery grounds hosting a spa well also took advantage of the fashion for taking the waters, in effect serving the purpose of spa gardens - a symbiotic relationship between the spa and garden nursery businesses. At Canterbury in Kent, visitors were principally drawn to the city for its Cathedral and religious importance as the centre of English Christianity, perhaps also for the annual race season. Although a chalybeate well existed and Canterbury Wells was marked on an 1810 map, it was the nurseryman William Masters FHS, (1796 – 1874), proprietor of 'Mr. Masters' Exotic Nursery' who took advantage of the waters and advertised from 1833 the Canterbury Nursery Grounds complete with spa well.

A mineral spring was also a desirable feature to add interest to an existing public park. Thus Hyde Park in London had a spa well, but of course this Royal Park can in no sense be considered a 'spa park'. Another instance is the mineral well in the eighteenth and nineteenth century pleasure grounds of the former Bishop's Palace at Bromley, later developed into an ornamental rockery by James Pulham in the 1860s. However, a spa fountain might well add to the interest of a newly planned development of a park such as Alexandra Park in Hastings, in which the Victorian landscape designer Robert Marnock incorporated a minor chalybeate well into its design in the 1880s.[33]

Social historians are interested in spas for their role in shaping the rise of leisure culture in Britain from the eighteenth century onwards. The importance of promenading cannot be over-emphasised:

> *But the last thrill of the cornet has just sounded from the band in the gallery, and the fashionable throng is sallying out of the saloon to spend an hour or two in visiting or walking, in observing and being observed, ere they sit down to their principal repast. This promenading is their daily bread...* [34]

The undoubted appeal of meeting up with other like-minded souls in an atmosphere of leisure and entertainment at a watering place benefitted the development and fitting out of spas from the earliest days, which in turn promoted this kind of tourism in Britain. A key reason for travel in ages past - just as now - was for health and diversion. Common ailments, the stresses of a busy (or boring) existence, excesses of a rich diet and the oppression of a dusty city in the Summer all took their toll and 'the cure' became a fashionable solution. Speculators capitalised on the existence of mineral springs to build up a base for tourism both within existing spa towns and at newly discovered local springs for speculative developments around a new spa. Prime examples of such developments include Pittville Park at the edge of Cheltenham, but also short-lived, unsuccessful ventures at Dorton Spa in Buckinghamshire and Hockley Spa in Essex.

Beyond the role of spas as centres of healing and recreation it is also fascinating how these places became venues for political and social intrigue.[35] The 'otherness' at spas where multiple public meeting spaces and socialising occasions were on offer was the ideal opportunity for 'transgressive behaviours' in an atmosphere of increased social freedoms at spas. This ditty based on Sydney Pleasure Gardens in Bath implies the joys of hidden retreats in the park:

In Sydney with your Venus,
The bustle of Bond-Street we'll shun,
Whilst pendant laburnums shall screen us
Awhile from the heat of the sun.[36]

Romantic liaisons were almost an expectation in the fashionable watering places, where alcoves and shrubbery will no doubt have been screening not just from the sun, but also from the gaze of passers-by.

In Britain, the earliest spas such as Tunbridge Wells and Bath benefitted from the relative difficulty of travel abroad that involved days of arduous coach travel as well as a sea crossing by sail. Whilst many health tourists would have preferred to travel to Europe, British spas sought to promote themselves as places with similar benefits without crossing the English Channel. Locations like Buxton in Derbyshire, Great Malvern in the west of England or the Welsh and

Scottish spas such as Llandrindod Wells and Strathpeffer had natural scenic advantage. A spa visitor to Malvern in 1775 wrote a recommendation, explaining that he was 'charmed with the situation of the Well-House and the beauty of the Country round it', it even 'exceeded [his] expectation'.[37] As late as the early 1900s, spas were vying with the appeal of travel to foreign resorts by emphasizing the benefits of pleasant prospects along with taking the waters without leaving the British shores. This comment in *Country Life* from 1906:

> *In the beautiful "Garden City" that Cheltenham has now become, with its lovely surrounding of hill and vale, it would indeed be far less of a hardship to make a 'cure' than to be sent off half across Europe*[38]

A number of observers have commented at various times in history on the declining fortunes of the British – mostly English – spa town in comparison with their more successful European neighbours. Indeed, as early as 1841, A. B. Granville had a whole section in his spa guide on 'Reasons of the recent disfavour of English mineral waters, and of the decline of most of the once -fashionable spas', stating that the minerals waters had been growing out of fashion for the last 30 years. He considered that Bath and Tunbridge had become nearly forgotten and Cheltenham and Leamington were being visited for reasons other than their healing waters. He put much of the decline down to misrepresentation or exaggeration of the benefit of the mineral water – which we, of course, know to be fact - but also identified failures of presentation of the facilities on offer. Muddy wells, surly 'dippers' (those dishing out cups of water), cold, dark pump rooms and draughty bathing facilities came up as definite repelling features which would drive health seekers to look elsewhere for a cure! On the other hand, those spas with attractive resources tended to charge visitors handsomely, which became unsustainable once the railway opened the way for affordable mass tourism.

Once the Napoleonic wars were over after 1815 and travel to the continent became safe again, the romanticism and fashionable trend of travelling abroad to 'take the waters' rather than patronising home-grown facilities took over. Those who could afford it did cross the channel to those places that A. B. Granville had been writing about in his book *The Spas of Germany, 1837*. Destinations such as Baden Baden (literally: Bath-Bath) and also fashionable towns in other European countries like Spa in Belgium (which gave these kinds of resorts their name), Vichy in France and others were popular for their scenic locations as well as the facilities. Those doctors and scientists of the time who made a study of the constituents of the waters and their supposed effects were generally agreed that most British spa waters were the equal if not superior to those found in other countries. (For modern mindsets this is an unsurprising finding in the light of understanding both of geology and medicine, that the actual waters were unlikely to be the major contributor to any health benefits experienced by the spa visitor.) Concern over the competition from foreign resorts even led to the *Health Resorts and Watering Places Act 1921*, which permitted local authorities at seabathing and spa resorts to raise fund from rates to pay for advertising of their attractions, but the trend for foreign holidays continued to grow.

Rapid railway expansion during Victorian times should have benefitted spa tourism by improving access to the spa towns. For a while, this was indeed the case in places like Buxton, Strathpeffer and Llandrindod Wells. But in the long run the opposite effect occurred, leaving spas to struggle in an increasingly competitive tourist market. The Great British public flocked to the seaside for fun and frolics, not for health cures. And the elite *clientèle* who maintained the tradition of spa visiting were now able to access the European 'watering places' with their warmer climates (and better casinos) more easily. There was a continuing momentum of spa recouperation breaks as part of the health systems in Europe where the 'cure' concept survived almost to the present day. Ultimately three factors were instrumental in the success of spas in mainland Europe: scenic value, social diversions and climate. Whilst Britain had scenic value in abundance and was able to offer suitable diversions in many locations, the climate presented the same problem as it does nowadays for anybody planning a holiday on home ground. In 1840, A. B. Granville wrote of the northern spas that 'Yorkshire possesses every thing desirable, except an Italian sky or a Lisbon climate'. At Bristol Hotwells, Granville analysed meteorological data which showed that in the six years from 1830 to 1835 there had been 1108 days of fair weather and no less than 1032 days of rain – without counting 51 days of snow. This moved the author to conclude: 'This is sufficiently dismal.'! Fifty years later, in a review of the 1895 report *Baths and Health Resorts of the United Kingdom,* the authors had to admit that 'while patriotically standing up for our British spas...such advantages as they possess are seriously handicapped by our climate'.[39]

After the world wars of the twentieth century, the case for people to choose spa holidays in Britain was unconvincing and 'taking the waters' declined, to be replaced with hotel-based 'hydro' bathing treatments and ultimately, the pampering, 'chill-out' experiences of our current age.

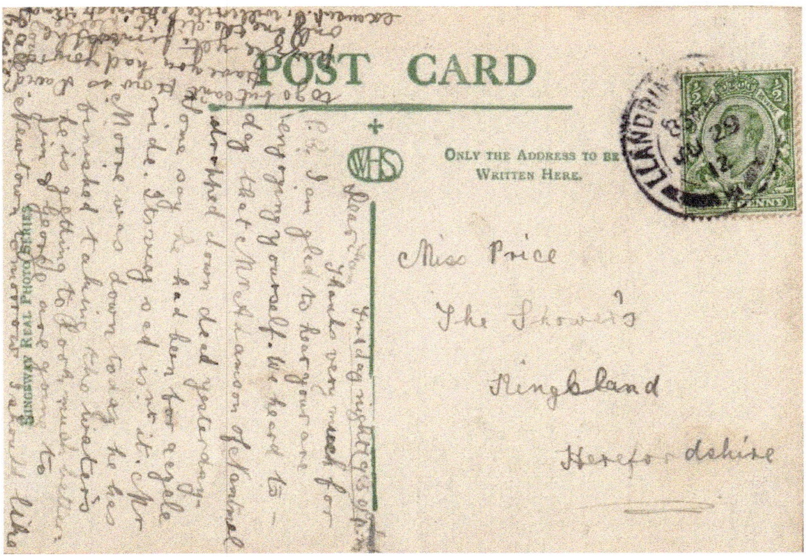

A postcard from Llandrindod Wells in 1912 still records visitors actively 'taking the waters', in 1912, but it was the last gasp (certainly for Mr. Adamson!) as far as British spas were concerned.

Auxiliaries to Healing

As a keen advocate of mineral waters being consumed or used for bathing at their place of discharge from the ground, our physician author A. B. Granville supported and promoted the concept of travel to specific spas. Critical to the choice of spa sojourn was obviously the type of waters available. The waters' constituents were scientifically analysed as to mineral and gas content and different waters were considered to effect different cures. In modern epidemiology, the wealth of anecdotal 'proof' would not be admitted as evidence of curative effects, but these were key factors for physicians recommending a specific spa.

However, there were additional aspects that needed to be considered for an effective cure, from quality of accommodation and food to the availability of entertainment venues. As visitors were encouraged to spend many weeks at the location, these also were important decisions to make, and A. B. Granville discussed at length those aspects of 'taking the waters' at a spa that he considered 'adjuncts' to the healing process:

> *There have been at all times persons...who have attributed the cure which results from drinking the [waters] to the many circumstances attending and accompanying their use.*

And he recounts these 'favourable circumstances' as:

- The journey
- A change of air
- A difference in the previous mode of living
- A release from laborious occupation
- Leaving behind every worry and anxiety of mind
- Gaiety and constant amusement amid agreeable society.

If this list looks suspiciously like the expectations we have for our own holiday travels, it is because we have become accustomed to the idea of holidays promoting well-being in mind and body. The stimulation of new surroundings, having fun with others and leaving our cares behind is the least we expect from a successful break to 'recharge the batteries.'

Granville describes his list of circumstances to support healing as 'auxiliaries to the powers and virtues of mineral waters' which can 'shorten the treatment, by rendering it more efficacious as well as more agreeable.' He notes the pleasure of

> *the contemplation of a beautiful country... or the sight of neatness, comfort, and order'- both on the journey to the spa and in the residence at a spa itself.*

One can therefore consider spa landscapes to be both an attraction to tourists and an 'auxiliary to healing.' Scenery and parks are no small part of the ambience in which to enjoy a healing holiday, but can designed landscape actually be part of the health cure? Hirschfeld's (page 9) pronouncement on this was that a good spa landscape provides easy and pleasant activity and 'diverts the soul from the feelings of bodily weakness' and 'enlivens the spirit with novel, delightful views'.[40]

Whilst the notion that a positive mindset promotes healing has been around for centuries, ongoing clinical research is now beginning to scientifically unravel the brain's control over healing processes. In the brain, just above the hypothalamus gland that controls immune responses sits a region called the insula, which is involved in processing emotion and bodily sensations. This proximity (of regions in the brain for emotions and immune response) is of interest, possibly linking emotional states to a person's healing. Work on understanding neurological pathways could soon map how both positive and negative stimuli could affect healing[41]. Considering spa landscapes, the positive stimuli that a planted outdoor environment could provide may well be significant.

ECOTHERAPY

The modern study of ecotherapy considers the effect of green space and planting for curative ends. It has spawned the concept of 'therapeutic gardens' which classify gardens into categories:

* healing
* enabling
* meditative
* rehabilitative
* restorative[42]

Common features of these gardens include what has been described as 'a profusion of human interaction possibilities.' Translated, this appears to mean:

- colourful and eye-drawing displays
- visually pleasing, touchable and fragrant plants
- recognisable comforting patterns
- accessible paths
- scheduled activities in the outdoor space

Accessible paths at Buxton Pavilion Gardens.

Despite the fact that such analysis was not articulated in quite this way when spa entrepreneurs laid out parks and gardens in the past, such features can be clearly identified in spa parks. A combination of functionality in the walks with attractive planting and views to promote that feeling of wellness in outdoor spaces is analogous to the modern work of the Therapeutic Landscapes Network, whose members work with hospitals and care homes for the same effect. The 'wellness' effect of green space therefore applies not only to the traditional spa but also in the design of contemporary spa facilities, many of which incorporate indoor or outdoor planting to enhance their therapeutic environment.

So, what is it that spa landscapes provide in particular for the health-seeking visitor? In the context of 'auxiliaries to healing' the following can be considered:

- Fresh air
- Exercise
- Access
- Communal gathering
- Shelter

Let us identify how these key elements for promoting healing were offered at British spas and specifically in their designed landscape.

FRESH AIR

For health and well-being, the outdoor green space at spas functions to support breathing in the 'fresh' air. A mistaken belief in 'miasma,' thought to be poisonous air causing disease, may have exaggerated the supposed benefits of fresh air. In any case, the overcrowding, heat and dust of a city such as London in the hot Summer may have been sufficient reason for fashionable folk to escape to a cleaner atmosphere. Lord Boyle, who visited Tunbridge Wells in the Summer of 1728 wrote that 'the foggy Air of London with the common Disappointments of Life urge our Return the following Year'.

The emphasis being on 'fresh' air, the most successful spas were located away from the big towns in rural locations. Even many of the spas of London required an outing to what was in effect countryside, such as the villages of Hampstead, or Epsom – or Islington:

> *No, give me fresh air, and Islington! – All the world shut up their houses in London at this time of year, and resort to the watering places.*[43]

Doctors recognised that getting away from urban stench was part of the treatment, something that Granville called 'ancillary' to the cure. Fresh air and 'the right kind of air' was crucial, particularly for those suffering from lung conditions, so spas needed to offer facilities for outdoor recreation as prescribed by the physicians. A 'Gentleman at Tunbridge Wells' wrote in 1777,

> *This place...deserves some little eulogium... The air of the place, and neighbouring hills, is remarkably salubrious; and the rural shady walks and murmuring streams ... forms it one of the most agreeable, as well as healthy, Summer retreats.*[44]

A hundred years later, the newly opened 'Spa Hotel' in the same resort claimed that 'it need hardly be said that the air is peculiarly pure and invigorating'[45] It also needs hardly be said that since British spas - particularly those in northern counties - were subject to typically cool weather, it was necessary to make a virtue of it by describing it as 'invigorating'. An enthusiast of the minor spa of Dinsdale-on-Tees in north-east England glowingly wrote of its 'healthy and *bracing* climate.' [46]

The author of an 1819 Guidebook of walks and excursions around Bristol's Hotwells spa, (peppered with poetic verses inspired by the locations) can have the last word on this topic of fresh air:

Airs of delight that soothe the aching sense,
Waters of health that through yon caverns glide.
Oh, kindly let your healing powers dispense,
And bring back feeble life's exhausted tide! [47]

EXERCISE

Exercise was an overt requirement at spas, prescribed by doctors as part of the treatment. A. B. Granville highlighted that,

> *After drinking the water, a little gentle exercise*
> *in the open air, if possible, should be taken, in*
> *order to effect the complete digestion of the water.*

There were physicians who even linked other medicines with bathing and exercise, such as *Dr Bodum's Medicinal Syrup* 'to be taken immediately after quitting the bath ... ten minutes walk after is necessary'.[48] The traditional spa routine is described in verse form in an 1816 publication:

At early dawn prepare to rise,
And if your health you really prize,
To drink the WATERS quick repair,
Then take a walk to breathe fresh air,
Hie thro' the fields – or promenade
Round Pump Rooms grand, or Colonnade...[49]

The provision of suitable walks and promenade opportunities close to the source of the mineral waters was therefore a key element of spa provision. It was generally considered that as the invalid started to feel stronger, they could undertake more walking and take the opportunity to ride or walk further afield in the surrounding countryside. At Strathpeffer Spa in the Highlands of Scotland, the owner of the Coul estate wrote in 1810,

The fresh air which circulates around the spring and the pleasantness of the county in which it is situated, by tempting individuals to walk abroad probably contributes more than the water to the restoration of their health.[50]

He obviously appreciated that both fresh air and exercise were important 'auxiliaries to healing.' This, taken together with the concept that aesthetic pleasure in landscape and the spiritual refreshment offered by contemplating nature also has an effect on well-being has led to the modern concept of 'therapeutic landscape'.[51] The UNESCO World Heritage site designation identifies 'Therapeutic Spa Landscape' as an element of the anatomy of the 'Great Spa Towns' of Europe:

A green or woodland environment used regularly for exercise as a contribution to the therapy of the cure, and for relaxation and enjoyment. Extensive networks of paths might be linked to particular medical conditions. Golf, tennis and horseracing were also provided. [52]

ACCESS

A. B. Granville discussed in detail the shortcomings of the newly created spa at the cliffs at Scarborough in the late 1830's at which 'the gravel walks...are sloppy in wet weather, and too loose and unconnected on their surface in exceedingly dry weather.' He went on, to emphasise,

This is no trifling consideration for visitors at Scarborough, who have to walk once if not twice a day in all weathers to and from the new Spa, for the purpose of drinking the waters, and who, even at other hours of the day, are disposed to make use of their privilege of meandering along the various paths on this cliff, which, when more thickly planted, will be a great resource to the invalids.

For the sickest of patients, allowance was made for helpers to collect water in bottles for those truly debilitated, but attendance at the pump house was desirable if at all possible. Bearing in mind that spas were visited by invalids hoping for cures, it was important that walks could be negotiated with relative ease, so dry, level, often gravelled surfaces were important for these avenues. Especially in towns where the three-wheeled Bath Chair – invented 1783 by John Dawson of Bath – was used to carry the sick from their lodgings, the walks in the parks were suitably wide with flat surfaces.

Vintage postcard of 'The Invalids Walk' at Victoria Park in Bath.

Where spas presented challenging slopes, such as Strathpeffer in the Highlands, it can be seen that the paths were constructed with handrails (page 99) and appropriately shallow steps for ease of climbing through the woodland spa park behind the pump room.

COMMUNAL GATHERING

Hirschfeld (page 9) wrote that spa parks require 'large spaces within planted areas, where whole parties can gather outdoors'.[53] At spa parks the facilities were not intended for the local population, but as an amenity for the health tourist. It was considered that the class of people visiting would appreciate the facilities and therefore subscription i.e. payment for entry was customary. This made the establishment of parks thoroughly commercial enterprises where large-scale events and performances were often laid on for guests. Pleasure Gardens such as those at the Cheltenham Montpellier Spa Gardens put on events drawing large crowds: festivals, balls and even spectacular feats such as the balloon ascent depicted below.

'Mr Hampton's Balloon and Parachute' October 3rd, 1838 at Cheltenham Montpellier Spa Gardens. Detail image G. Howe, Published by H. Davies, Cheltenham. ©The Board of Trustees of the Science Museum, London.

The benefits derived from the spending power of health visitors made the funding and establishment or improvement of spa parks a better proposition than parks that relied on schemes associated philanthropy or public subscription – as long as the visitors kept coming.

SHELTER

The tree-lined avenues at historic spas are frequently mentioned in the guidebooks and pamphlets as opportunities for 'shady' walks, giving the impression that a cool retreat from the glaring sun was an important consideration. Whilst there may have been many a hot Summer's day at British spas, it is as likely that wind and rain were the elements from which shelter was needed!

A colonnaded walkway at the Royal Pump Room and Baths at Leamington Spa in Warwickshire provides shelter with floral adornment.

This is certainly evidenced by the fact that covered colonnades and ambulatories were routinely incorporated into spa architecture. Extensive walks were available under cover at the larger spas (pages 56 and 62) and the colonnades were often prettily adorned with nearby-by flowering bushes and container planting.

Bearing in mind that drinking copious amounts of water has inevitable physical effects, a further consideration for those drinking the prescribed quantities of water at the wells was the requirement to 'step out' from the company for necessary relief. An account from 1632 of taking the waters at Epsom describes that 'the greater part of those that drinke of it, are purged by stoole, and some by vomit, as well as by urine'.[54] Although eighteenth century folks were rather less sensitive about the bodily functions than we are today, this universal necessity surely would have required at least some visual shelter. The Dutch visitor, William Schellinks, who came to Epsom later that century amusingly portrays how on the open heathland surrounding the well 'gentlemen and ladies have their separate meeting places, putting down sentinels in the shrub in every direction'.

Perhaps tree-lined avenues at spas helped to delineate the public walk from surrounding zones into which visitors might 'retreat' as required. Obviously, there were also likely to have been designated areas, although the record on this topic is understandably sparse. Our physician, A. B. Granville does record 'a certain unbecoming arrangement' at the Montpellier Spa at Cheltenham, which gives an insight into public facilities at tourist spots in the first half of the nineteenth century:

> *... the indecency here is even more glaring, as the entrance to the objectionable retreats is close to the public walk, and no bibber [drinker] of mineral water detaching himself from the throng of promenaders of both sexes, to dart down the trellised footpath leading to those retreats, can fail to have the eyes of everyone upon him. This demands instant reform.*

It will be noted that the functions discussed would be desirable for any park that hosts public gathering and the author Kaspar's plea to regard spa parks as a separate category (page 14) may be overstated, at least in Britain.

Elements of Designed Spa Landscapes

A consideration of spa parks and gardens by the German garden historian, Katrin Schulze has enumerated recognisable elements within a suite of designed landscape at spas:

- *Brunnenallee* (Well-walk/promenade)
- Promenades and path systems
- Use of shrubs and ornamental planting
- Built elements
- Play and sport facilities

These elements would be provided with intent that focuses on providing amenities for health-seeking visitors. Inevitably, the fashion of the age would modify the actual design. Early spas would typically provide a straight, tree-lined walk, whilst Victorian spa parks offered islands of colourful bedding displays in lawns with a network of sinuous paths. Modern spa hotels are providing restful nooks and crannies surrounded by hedges in which to relax in privacy. The common aspect is the consideration to provide the landscape linked to the therapy. 'The Queen of Spas – i.e., Bath, certainly included 'Beautiful Parks' in its advertising copy for magazines such as *Country Life* (for further details you would need to 'Write to Baths' Manager for Illustrated work, *Bath as a Health Resort.* Price One Shilling).

Schulze admits that 'it is not possible to def ne the spa garden or spa park as a special type in garden history,' but:

> *'What is surely characteristic ...is the close association between the function and use of the spa park (as a whole or as individual elements) and its creation.*[55]

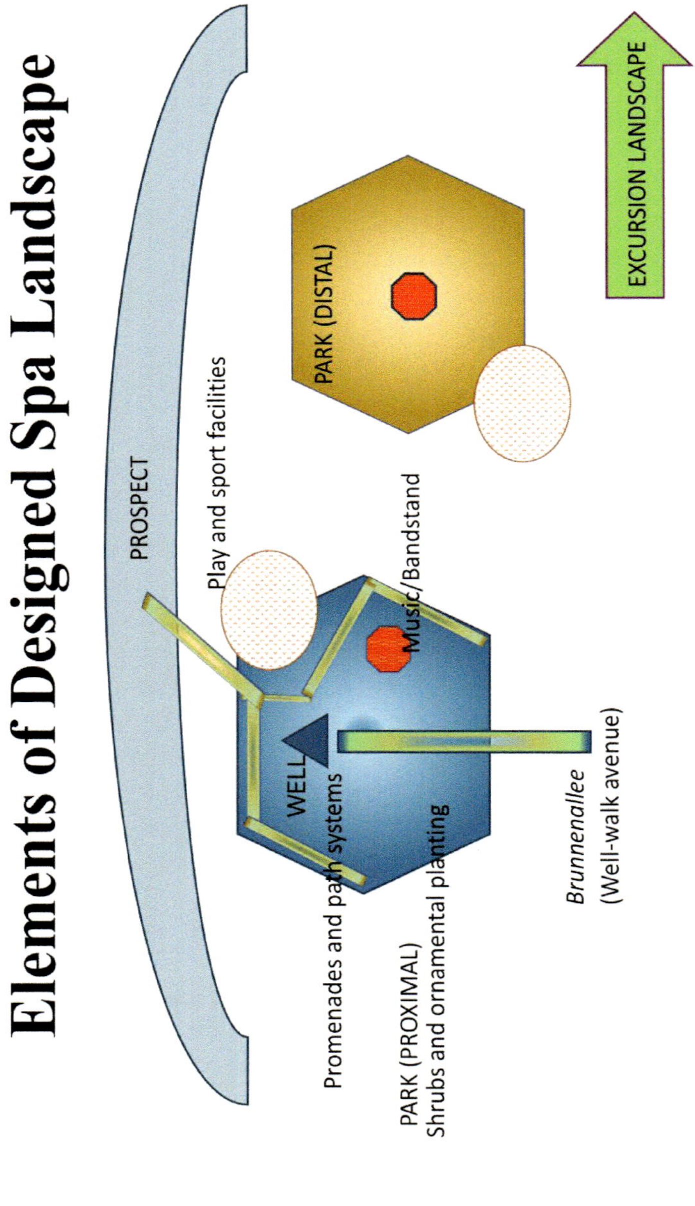
Elements of Designed Spa Landscape
PROSPECT
Play and sport facilities
Music/Bandstand
WELL
Promenades and path systems
PARK (PROXIMAL)
Shrubs and ornamental planting
Brunnenallee
(Well-walk avenue)
PARK (DISTAL)
EXCURSION LANDSCAPE

BRUNNENALLEE: WELL WALKS AND PROMENADES

Promenades in the formal gardens of the seventeenth century – the French term *allées* was used – proved to be an optimal arrangement for communal gathering at grand country houses as well as providing pleasing vistas outwards from a grand building. Tree-lined avenues likewise have been a feature of landed estates in Britain since the seventeenth century, 'divers trees, enrang'd in even rankes' being mentioned as early as 1596 in Edmund Spenser's book *The Faerie Queene*. Young heirs of country estates experienced the continental fashion for avenues when on their 'Grand Tour' of Europe and rows of tree trunks recalled the architectural grandeur of the colonnades of classical antiquity.

At spas, avenues developed where perhaps otherwise no such feature would have been created. The 'Well Walks' were designed with a focus *towards* a structure of interest to access mineral waters – a well or pump house or pavilion serving spa waters. Early spa-based walks would typically be straight, wide, and tree-lined to best meet the requirement for a considerable number of people accessing the waters at around the same time, usually between 5.00 and 8.00 in the morning.

OTHER PROMENADES AND PATH SYSTEMS

Other pathways are, at times, difficult to separate from the 'Well Walk,' but they would be further 'walks' that extended the opportunity for exercise – and social promenading - after taking the waters. Hirschfeld also advised:

> *Wild footpaths for walking, riding or driving upon for the guests of the wells ... for longer and further exercise and diversion.*[56]

Not all avenues were strictly straight by any means. From the eighteenth century the geometric formality of gardens had relaxed into the 'Picturesque' style which accommodated curves and irregularities in an artful fashion to present an idealised representation of natural beauty. From the nineteenth century, walks and avenues of the 'Gardenesque' style skirted around neat lawns, past colourfully planted island beds, grouped shrubs and spaced out specimen trees. At spa parks like Beulah and Buxton, for example, gravel walks in curves and serpentine lines were key elements of the park design.

At the smaller spas, where visitor numbers were less likely to overcrowd walkways, narrower and winding pathways offered sufficient promenading opportunities. Strathpeffer Spa grounds, Llandrindod Wells' Rock Park and St Ann's Well Gardens at Hove all supported abundant pathways on relatively restricted, sloping ground adjacent to pump houses. Such intersecting paths over the hilly topography provided 'refreshing foliage and shady walks' specifically for exercise for the spa visitors. In addition, destinations of interest were often incorporated such as viewpoints, ponds, bridges and picnic fields – at St Ann's Well, Hove - even a hermit's cave.[57]

1930s postcard St Ann's Well Gardens, Hove with a hermit's cave.

SHRUBS AND ORNAMENTAL PLANTING

Colourful embellishment planting is an important part of the attractiveness of spa architecture and accommocation. In 1777, our aforementioned 'Gentleman of Tunbridge' wrote ,

> *The lodging houses (almost every one of which, enjoys either an ornamental clump of trees before it, or a pleasure garden around; and some, the pleasing variety of both) seem notwithstanding, to be striving for rivalry.*[58]

At Leamington Spa, Hopton's Boarding House advertised in the newspaper that they, '...earnestly solicit the support of the spa visitants ...' and claimed that 'The situation is particularly pleasant, being situated in the centre of a flower garden...'.[59] By the twentieth century, the poet John Betjeman was still enamoured with planting in spa towns, as expressed in his contribution to the 1945 '*Britain in Pictures*' book. Describing Cheltenham, he particularly admires the 'town in the Regency manner' with 'crescents and avenues...shaded by trees and lightened with flowering shrubs.'[60]

Vintage postcard: Colourful bedding in front of Harrogate Royal Baths.

Beyond the embellishments at private boarding houses, it was found that at least 40 British spas had gardens or parks planted specifically to be associated with their mineral water source. Clearly, the designs of these gardens varied greatly from simple planting flanking a water tap to the whole-scale design of parks by well-known landscapers. Early spa gardens of the seventeenth and eighteenth century, if present, usually provided adornment with flowers and bushes close to a well or pump house, but on a small scale as most of the landscape was still essentially rural and the natural landscape was visually accessible.

Design and choice of plants at spas was influenced by the perceived effect on the visiting spa public rather than (as is usual in garden history), a display of private means or for the personal expression of a landowner. In parks, prevailing garden fashions were incorporated, but modified to appeal to a wider audience. Tidiness and order in both the lay-out and planting have been key features of all parks, but in spa parks the maintenance of this appearance was particularly important. Although the garden fashions of the times are represented, most design is likely to have catered to conservative tastes rather than novel approaches. Typically opulent Victorian carpet bedding displays (pages 14, 43, 79) have, for example, maintained their popularity right into the current century and are one of the features of spa parks that still draw admiration from visitors.

21st century spa park planting in Valley Gardens, Harrogate (2023).

In the 1920s, a novel idea at Royal Leamington Spa was the provision of floral decorations in the streets, a scheme that 'attracted widespread interest and admiration'. The intention was to popularise the spa town at a time when spa cures in Britain were on the wane. It was reported that 'Americans were frequently making inquiries and asking for photographs of a scheme regarded as being unique.'[61] If true, we have a spa town to thank for instigating the colourful floral displays that are now a feature of welcoming High Streets throughout Britain.

Although floral displays in parks and at horticultural events had been common since Victorian times, adorning streets with tubs and hanging baskets for purposes other than special occasions such as a royal visit was new. It was not until 1963 that the concept of judging flower displays in towns and villages was rolled out as a 'Britain in Bloom' pilot, the inaugural competition being won by the spa town of Bath.[62] It was reported that 'shopping centres and outskirts were equally attractive' and that 'In the town centre every lamp-post, pillar-box and telephone kiosk had its floral decorations'. Interestingly, one of the prizes was a trip to the spa town of Harrogate![63]

1970s postcard: Bath in Bloom.

BUILT ELEMENTS

For most minor spas, modest shelters covering a spring served as the pumphouse at which one might queue to have the healing water dispensed into the drinking vessel. Wooden roofed structures covering a well or spring of perhaps 2-3m diameter were common, perhaps with a low brick wall surround or simple plank edging.

Rock Park, Llandrindod Wells.

Valley Gardens, Harrogate.

At the other extreme, the large fashionable spa parks incorporated design elements such as imposing entrance gates, fountains, steps and terraces and kiosks in keeping with the garden and landscape fashions of the day. The pump room would be the significant feature, with a wide variety of structures that varied from rustic huts to grand edifices in classical style. The pumproom in Pittville Park (pages 70, 71) on the outskirts of Cheltenham was certainly grand and still stands proudly in this former spa park.

Imposing architecture was, however, not confined to the larger urban spas; smaller speculative establishments attempted to attract interest by employing well-known architects to create an impressive centrepiece for a spa. At Dorton Spa (page 92), in a very rural location five miles southwest of the Waddesdon Manor estate, an elegant pump room was designed by architect James Hakewill (1778 – 1843) as centrepiece in specially created pleasure grounds. A building plan for the pump room shows a design for exits at each corner from a ballroom and billiard and reading rooms, as well as the central entrance, making access to manicured gardens part of the design.[64]

At Beulah Spa in Norwood, South London (page 90) the great architect Decimus Burton (who was also responsible for the Calverley estate at Tunbridge Wells) designed the well house as a wigwam construction and a reading room that displayed an extraordinary amalgamation of a classical roof spire and rustic colonnade, the branch-like beams reflecting the picturesque fashion for rustic elements.[65]

An equally unusual construction of 1862 at Tenbury Wells was designed by James Cranston, a Victorian architect who had previously published a book of designs for garden greenhouses. The 'Chinese Gothic' pump rooms have a hipped roof covered by wrought iron sheets. It was one of the earliest examples of prefabrication, the sheets assembled on site.[66]

Pump Rooms based on a greenhouse design at Tenbury Wells.

ACTIVITIES IN THE PARK

Games - such as bowls - and music have definitely played an important part in spas acting as diversion for visitors who stayed often for many weeks. When Celia Fiennes visited 'The Wells' in 1697, she described how 'the Gentlemen bowle [and] the Ladies dance or walke in the green in the afternoons'.[67]However, the rise of sport facilities and bandstands in public parks generally means that it is often hard to link their provision specifically to spa status.

Historic bowling green behind the Pump Room at Strathpeffer.

PLAY AND SPORTS FACILITIES

In the later nineteenth century, public parks were providing opportunities for recreational sports such as cricket, archery and tennis, and so sport began to feature in spa and other resort parks as well. At Buxton, the physician Henry Robertson recorded that 'level lawns and asphalted spaces will supply any wants that may be indicated from time to time for the out-door games that may happen to be in vogue.'[68] At Strathpeffer spa in the Highlands, a well-known golf course became a key attraction.

OUTDOOR MUSIC & DANCING

The tradition of musical entertainment at spas goes back to the seventeenth century. In Bath, for example, mention had been made in 1651 of 'a Band of Musick consisting of viols and oboes performed in the Orange Grove.' [69] In Northamptonshire at King's Sutton Spa that intrepid traveller and spa enthusiast Celia Fiennes visited in 1694. Her writings recorded a

> *'Fine Gravell Walke that is between two high cutt hedges where there is a room for the musick and Roome for the Company beside a Private Walkes'.*[70]

Around 1700, Metellus, a 'gentleman-commoner of Christchurch' wrote of his visit to Tunbridge Wells where he noted the tree-lined shady walk of what is now known as 'The Pantiles' and declared that 'Midst of the Trees Apollo has a Quire'; music being as essential to the cure as the waters themselves.[71]

Vintage postcard : Location of the Orchestral Balcony at Tunbridge Wells.

Dancing also could be enjoyed in the outdoors. Another visitor to Tunbridge Wells in the early 1700s confirms this leisure activity:

> *As soon as the evening comes everyone quits his little palace to assemble on the bowling green, where, in the open air, those who choose dance upon a turf more soft and smooth than the finest carpet in the world'*[72]

Since entertainment was the key attraction of pleasure gardens, those spas with 'Vauxhall' type gardens will have had space or structures provided for music and play. For example, in the 1840s the Montpellier Spa at Cheltenham provided musical promenades in the evening four times a week during the Summer either in the pump room and promenade walk, which were brilliantly lighted up for the occasion, or in the adjoining gardens.

Many of the larger spa parks had a bandstand at some time in their history with the earliest erected in 1864 at Montpellier Gardens in Cheltenham.[73] However, this date falls within fifty years of effectively the end of British spa activity and therefore British bandstands are more a feature of the municipal public park movement than – as in other European spas – a direct function of spa entertainment.

Pump Room Gardens, Leamington c. 1910 - in 'Lenton & Sons Grand Photographic View Album of Leamington Spa & Shakespeare Country'.

UNIMPROVED LANDSCAPE

In researching landscapes associated with spa towns and spa wells, it becomes obvious that there are large numbers of former spas where the landscape was not deliberately 'improved' to create gardens. There may have been 'walks', opportunities to take exercise in a rural or urban setting, but no deliberate laying out and planting of avenues, lawns, shrubbery or flowerbeds. Examples of landscapes identified under this category were undeveloped meadows, sea-side promenades, riverside paths and other locations with scenic vistas. One example might be the Brow Well on the Solway Firth in Dumfries and Galloway at which the Scottish poet Robert Burns took the chalybeate waters just before his death in 1796. As a minor 'poor man's spa' there was no adornment as such, but the coastal location was a scenic location at which to spend time.

EXCURSION LANDSCAPES

Visiting a spa for 'the cure' was, of course, also a kind of holiday, so visitors who were well enough expected to go on outings in the surrounding district. A. B. Granville wrote copious pages on the delights of such excursions. Walking, on horseback or in carriages, it was popular to visit local landmarks but also country houses and their parks and gardens. Thus excursion landscapes like Prior Park at Bath or Studley Royal within easy reach of Harrogate would have been typical destinations. Spa guests, usually in social groups, would have departed after their morning 'duty' of taking the water. Smaller spas particularly benefitted from the existence of excursion landscapes, as their own offerings of entertainment were limited. For instance, A. B. Granville made particular mention of 'the beautiful drives, and many noblemen's seats, gardens and parks' for Ashby-de-la-Zouche in Lincolnshire, including special mention of Calke Abbey and Melbourne Hall.

HOTEL SPA GARDENS

The distinction between spa parks and spa gardens may be simply one of size, but where individual establishments had planted areas – notably spa-based tea gardens and spa hotels, they could be considered as spa gardens rather than parks as they were provided specifically for their spa clientele. Some supplied (and charged for) water from an existing - or opportunely discovered! - well to drink, but many were hotels that provided facilities for bathing in the mineral water. The issue, as discussed by A. B. Granville is the practical one of taking the baths in suitable surroundings whilst remaining close enough to the source of the water not to lose its efficacy (as was thought at the time).

> *Mineral baths should be near or contiguous to the dwellings of the invalids...especially where the climate is so variable, and the ground liable to get wet.*[74]

Later in the nineteenth century the so-called 'water cure' of the austere hydropathic movement of Dr Vincent Priessnitz spawned a multitude of 'Hydro' hotels, forerunners of our modern spa hotels, whose gardens in the main reflect the usual embellishment of hotel grounds.

LANDSCAPING ELEMENTS AT MODERN SPAS

Relaxation, 'pampering' is the modern understanding of spa treatment, but the concept of visiting a destination away from home is maintained. The ambience is carefully designed for functionality and appeal, and often modern spas include an outdoor setting as part of the facilities. The newly named 'Spa at Carden' is set within the 1,000-acre grounds of Carden Park in Cheshire. In 2019 it unveiled its brand new, specifically designed spa park, described as a multi-sensory wellbeing garden, the stated aim to 'elevate guest wellbeing by immersing them in nature. A centrally positioned kiosk dispensing champagne (other drinks are available) has echoes of a pump room, whilst winding paths with a surface suitable for barefoot walking serve as connecting walkways between sheltered spaces for private relaxation in pools or on deckchairs.

The planting at the luxurious Cheshire spa development was designed by Sarah Rycroft, who has a track-record for designing therapeutic gardens. The outdoor space incorporates a hydrotherapy pool, sun terrace, an outdoor sauna and fire pits. Low hedges provide the desired privacy for small groups of users and planting in repeated, subdues tones using a variety of shapes and textures gives a calming effect to support the 'chill out' atmosphere these modern spas cultivate for spa pampering. Crucially, the concept of the garden as an auxiliary to well-being is maintained. Leaving aside the fact that unwanted 'nature' - such as bees - were discouraged by avoiding plants attractive to the insects, the recognition of the contribution of nature and landscaping is equivalent to spa park design of the past. Even indoors, the concept of 'outside-in' incorporates an indoor 'living wall' of green plants as guests enter the spa building.

Modern spa park © The Spa at Carden - Cheshire's Country Estate.

The Green Legacy of British Spas

The landscape features that developed as 'auxiliaries to healing' at British spas are still with us to some extent, part of the rich pattern of designed public green space in this country. As well as a legacy of fine public parks, the spa towns were also pioneers in the regular use of ornamental planting in public places to enrich the visitor experience. In fact, in 2021 government funding was made available for councils to 'boost the look and feel of their high streets by investing in street planting, parks, green spaces and seating areas to make high streets as beautiful and welcoming as possible', a sentiment which echoes exactly what spa towns were aiming to achieve.[75]

The recognition of spa-related landscape is however just beginning to be appreciated. In recent years, authorities and interest groups in individual former spa locations have revived the history of their locality and have sought to restore - or at least publicise - elements of their spa heritage. Much of this inevitably focuses on buildings or structures, but the setting of these are most frequently the former spa parks in which they were located. There are places where funds have been accessed for physical improvements and promotion of a former spa park and information on interpretation boards, or leaflets, or indeed guided tours are increasingly available to the public. In 2021, this study identified at least twenty-four locations where the spa heritage of the park or garden was publicised or at least mentioned on-site.

In England, some spa parks are fully described by the Historic England Register of Parks and Gardens of Special Historic Interest, but without particular reference to their original intent to support health tourism. Likewise in Wales, CADW's register of historic landscapes does not even list Rock Park, the fine spa park at Llandrindod Wells (page 97), although its pavilion at the Spa Treatment Centre is a designated historic asset.[76] Thus in England and Wales, spa landscapes appear to lack the specific recognition due to them. In Scotland, there is more recognition of spa-specific landscape as demonstrated by

Strathpeffer. Here the Pavilion Spa Garden is a 'designated landscape' that is acknowledged by Historic Environment Scotland as

> *an unusual example of pleasure grounds laid out as part of a nineteenth-century health resort [...] interesting for their associated social and economic history.*[77]

Local history societies are often actively pursuing research and recording spa-related history in their area. Notably at South Norwood in the outskirts of London, The Beulah Spa History Project has ensured that the extraordinary story of this spa and its designer park is remembered through publications, talks and a commemorative ground level granite marker stone – 'The Beulah Stone'.[78]

Recent resurgence of interest in parks generally is however spawning more research into public green space and the Gardens Trust have run a series of internet 'Zoom' talks on Spa Landscapes in 2022, with speakers elucidating the fascinating histories of spa parks and gardens at Bath, Buxton, Royal Leamington Spa and the lost spa of Dorton in Buckinghamshire. Such recognition of the heritage of individual former spa parks could perhaps become an important driver for the care and enhancement of some of these public parks. A prime example of impressive park restoration is the recent (2020) refurbishment of Bath's historic pleasure garden, Sydney Gardens, which attracted £2.7M of National Lottery funding for major restoration.

The association between gardens and the 'leisured classes' has always been strong, but in spas the landscaping moved away from the individual whims of private landowners or patrons by considering and incorporating features of purpose and taste for a more general visiting public. It falls to an antiquarian writing in 1837 about a former spa well in a town that did *not* provide spa landscaping to summarise the purpose of such planting:

'If at this juncture a handsome pump room had been erected embellished in front we will say by an enriched colonnade of pillars surmounted by a dome and the contiguous grounds laid out in walks in a tasteful manner in order to blend utility with comfort and pleasure an attraction would have been presented to entice company to Wellingborough Red Well.'[79]

Vintage postcard: Colonnades in Harrogate Valley Gardens.

In his 1976 book *The Spas of England*, author Peter Neville Hawins avoided the mention of parks or gardens almost entirely. Yet he concluded that remaining former spa towns were, in effect, 'garden towns ... in many ways ready-built leisure parks' in which leisure and recreation had begun to return to an urbanised setting after centuries of 'flight from the towns'.

The spas have proved that urbanisation need not be a concrete jungle... especially when we realise that our dwindling open countryside cannot last forever.[80]

'Leafy Leamington': tree-lined avenue in a classic spa resort.

It is clear that the underlying fabric of successful spas was suffused with landscape – vistas and tree-lined avenues, lawns and flower beds, benches and bandstands. The provision of parks and avenues at these historic places of healing - justified on economic grounds – to some extent led the way in demonstrating the desirability of integrating green spaces in towns to enhance the living space.

PART II
SOME BRITISH SPA LANDSCAPES

Well Walks and Spa Promenades

As one of the key spa landscape characteristics (especially of early spas) the well walk or specific spa promenades make an appearance in at least 33 former spas, of which just over half retain vestiges of this feature (page 114). Examples in chronological order:

1703 The Walkes (Pantiles), Tunbridge Wells

1707 Avenue to the well, Epsom Spa

1740s Well Walk, Hampstead Spa, London NW3

C18 Riverside & Lovers Walks, Matlock Bath

1740 The Old Well Walk, Cheltenham Spa

1740s Approach from river to Park Wells, Powys

C18 Avenue on Walk Field, Witham Spa, Essex

1770s The Parade, Bristol Hotwells

1780s Riverside Promenade, Southampton

1820s The Slopes, Buxton

1830 Pump Rooms walk, Pittville Spa, Cheltenham

1830 Upper and Lower Walks in Valley Gardens, Harrogate

1831 Victoria's Walk to St Ann's Well, Malvern

1875 Linden Walk, Leamington Spa

The Pantiles at Tunbridge Wells: the original 'Well Walk'
... a row of trees, which has a gallery in the centre for music. [81]

Tunbridge Wells
The Walkes (Pantiles) 1638

Tunbridge Wells – which might be dismissed as 'only a provincial town'[82] – provides the prime example where a tree-lined 'Walke' was unequivocally linked to its status as a spa. After the 1606 discovery of a chalybeate spring in open country near Tonbridge, a spa facility was built up with the earliest well walk in Britain created in 1638, funded by public subscription.

Lime trees still delineate the parade which is known as 'The Pantiles' - paving stones that were laid to prevent further accidents, as Queen Anne's son the Duke of Gloucester had suffered injury on the irregular surface on a royal spa visit in 1698. Parallel Upper and Lower Walks on either side of the single line of trees provided opportunity for traders to offer their wares as the ladies and gentlemen frequented the springs:

Under the Fragrant Leaves
of Yonder Trees
You ready gather'd find
fresh strawberries;
With Odoriferous Rasps;
beneath that Tree
Shaded with Poplar Leaves,
you Cherries see...[83]

Even now the tree trunks on the one side are mirrored by the classical colonnades of the shops on the opposite side, a covered walkway sheltering the shops and cafes which became part of the spa experience. The tradition of music on 'The Walkes' also continues with a modernised version of the bandstand providing live music under the trees in the Summer months.

Epsom Spa, Surrey
Tree-lined Avenue to the well, 1707

Even in thoroughly rural locations, the well walk was a designed feature as at Epsom in Surrey. Originally no more than heath scrubland, the spa was described in the seventeenth century as a simple well house sitting among 'shrub in every direction'. Old maps show the final access from the road as a tree-lined avenue, the planting undertaken as part of the enhancement of provision by a new Lord of the Manor in the early eighteenth century.[84]

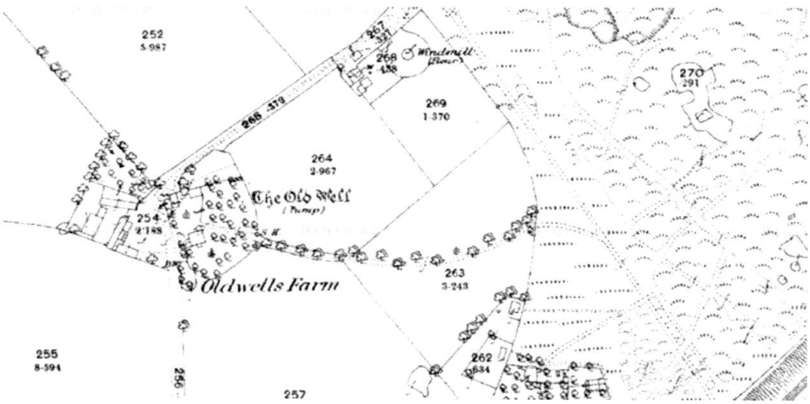

Hampstead Spa, London NW3
Well Walk 1740s

The first edition Ordnance Survey maps of the 1860s shows a straight road lined with trees for access and promenade specifically for visitors to the former chalybeate spring of Hampstead Spa from the early eighteenth century. The typical spa landscape has been lost to suburban housing; the name of the road remains as 'Well Walk'.

Matlock Bath, Derbyshire
Riverside & Lovers Walks 1740s

This scenic valley in the Peak District enjoyed notable visitors to its spa, including the poet Byron, who compared it with Alpine Switzerland, and the artist JMW Turner. The locality benefitted for centuries from the extensive woodland walk alongside the river – according to an interpretation board, mentioned in 1742, the 'oldest continuously used pleasure ground in the country'. *Matlock* Bath, the poem by John Betjeman mentions 'pleasant acres/Of intersecting LOVERS' WALKS' - These walks (listed grade II*) along the bank of the river Derwent are reached by a bridge close to the pump room. Additional ornamentation and landscaping were carried out, increasing the range and achieving dramatic views from the top of the cliff.

Matlock Bath, Lovers' Walk detail. Photo © Mike Faherty (cc-by-sa/2.0).

Cheltenham Spa
The Old Well Walk 1740

The visual grandeur of large avenues led to the creation of impressive tree-lined approaches to the front of country houses to proclaim status and style. At Oakley Grove (now Cirencester Park) Allen Bathurst, 1st Earl Bathurst (1684–1775) created a landscaped park with 'Broad Avenue' in 1710, a feature appreciated by King George III and his Queen and princesses when they themselves were visiting Cheltenham Spa in 1788.[85] It is then no surprise that in Cheltenham such grandeur had been emulated to attract the best in society - a public subscription walk for the spa clientele. In 1740 the now lost 'Well Walk' was created there with a double row of 143 trees for an impressive distance of about 360 metres.

> *There is so imposing an air of formality and antiquity in that beautiful avenue of elm-trees called "The Old Well Walk" which forms the principal feature of the "Original Spa", that one is tempted to select it [...]for a morning "constitutional walk".*[86]

'Old Well Walk, Cheltenham , George Childs Lithograph c. 1842.

Park Wells, Powys, Wales
Approach from river 1740s

This diminutive spa near Builth Wells with both a pump room and a bath house retains a grand tree-lined approach from the days when spa clientele used to approach the facility from the river.

A tree-lined approach to rural Park Wells.

Witham Spa, Essex
Avenue on Walk Field, C18

In this rural village within easy reach of London, a straight, tree-lined avenue was specifically created to access its spa well on a piece of ground called 'Walk Field', in keeping with the facilities that spa visitors expected. This feature still shows up on a map of 1870 but is now subsumed into a housing estate. The former well avenue, which led from the road to the water source actually remains as a tree-lined footpath at the rear of new housing.

Bristol Hotwells
The Parade, 1770s

> *If you to Bristol should repair,*
> *To drink the Waters, breathe the air*
> *To Pump-room gently walk or ride*
> *And in the Water's use confide ...*
> *Then softly tread the neat Parade*
> *Or stroll beneath the Colonnade...*

A river-side well-walk parade leading to Hotwells pump-house was a grand addition to the spectacular vista of the Avon Gorge. This attractive avenue in a truly sublime (as in awe-inspiring according to Picturesque theory) was described in 1800:

> *For those who prefer walking after taking the Water, [...] a gravelled parade about 600 feet long, shaded with trees, by the side of the river...*[87]

The spa had long been abandoned when in 1926 a new road to to Bristol's port at Avonmouth known as 'The Portway' completely obliterated the former Hotwells Parade.

'The Rownham-Hotwells ferry' by S. H. Grimm c.1780 -90, showing Avon Gorge with Hotwells on the right bank. Reproduced with permission from Somerset Archaeological and Natural History Society (SANHS) Grimm collection.

Southampton
Riverside Promenade, c.1780s

Southampton was regarded as a fine spa in the eighteenth century, with the benefit of views across the River Test to the New Forest. It had the distinction of royal patronage, visited by King George III and his entourage in 1789.

A tree-lined riverside promenade led up to pump rooms through attractive gardens. Any remnants of these from the nineteenth century were completely eradicated in the harbour industrial developments of the twentieth century, becoming the site of the Dunlop tyre manufacturers.

Southampton Pump Room in the nineteenth century – destination at the end of the Test riverside walk .

A 'Spa Road' still exists surrounded by offices, but Southampton spa landscaping was completely superseded by nineteenth century planned public parks in a different location east of the former spa.

Buxton
The Slopes 1820s

Mineral springs known from medieval times drew visitors to this rather remote location on the Derbyshire uplands. By the late eighteenth century, the fifth Duke of Devonshire began to promote the spa, building a grand Crescent. It is thought that the architect Jeffry Wyatville (famous for work on Longleat and Windsor Castle) was commissioned to carry out the original work to the sloping green space in front of it. Described on an 1828 drawing as 'The Terraces opposite to the Crescent', they incorporated the original St Ann's Well. Later alterations by Joseph Paxton (of Crystal Palace fame) in the 1850s extended the walkways across this green area to create a whole network of paths over the slopes, which still remain today.

It is fascinating that as part of the visual landscaping improvements, the Duke of Devonshire had part of the hillside in the distance planted with trees to create a more picturesque vista in what had been considered a rather bleak landscape.[88]

Vintage postcard featuring a network of promenade paths on 'The Slopes'.

Pittville Spa, Cheltenham
Approach Avenue to Pump Rooms, 1830

Later spa parks tended to have shorter well walks. Plans show that the Pittville Pump Rooms, opened in 1825, still maintained the tradition of a Well Walk by means of a central tree and shrub-lined avenue leading to the entrance of the grand building (page 71).

1970s postcard: Pittville Spa Park showing the open space created by removal of the former well walk in front of the Pump Room building.

This central pathway was later dispensed with, leaving open lawns. A Pittville Park Management Plan acknowledges that although the layout of the paths in the park as whole still closely reflects Pitt's original plan, the main difference is that the sweeping gravel path between the Pump Room and the upper lake was removed and grassed over when acquired by the Council in 1890.[89] This obviously reflects the change in the purpose of the park, now acting as a public 'People's Park'. The park nevertheless retains its former elegance, the Pump Room now popular as a wedding venue (2024).

1906 Postcard of Pittville Spa showing the central well walk avenue.

Harrogate Spa, North Yorkshire
Upper and Lower Walks in Valley Gardens, 1830

The northern spa of Harrogate had the benefit of a number of diverse kinds of wells, connected by paths. This network of promenades is a particular feature of this spa park, the magnificent Valley Gardens, which were eventually laid out much later to commemorate Queen Victoria's Golden Jubilee in 1887. The design of the gardens was highly influenced by the valley topography and 'Upper' and 'Lower' walks developed, the former under the protective covering of a colonnaded roofed walkway (page 56) which added appropriate grandeur to the promenade in the park.

The heritage of the park is proudly promoted, with restored wells and signage explaining how the variety of mineral springs were accessed for 'taking the waters' in times gone by.

Valley Gardens spa park with a network of promenades linking the wells.

Great Malvern, Worcestershire
Victoria's Walk to St Ann's Well, 1831

On the slopes of the Malvern Hills, a formidable zig-zag path leading up the steep hill to a spring at St Ann's Well was opened by Princess Victoria. Then – as now – visitors would be pleased to ascend for both the views and the sense of achievement!

Detail from 1910s Postcard: Great Malvern 'Zigzag' path to St Ann's Well.

Leamington Spa, Warwickshire
Linden Walk, Pump Room Gardens 1875

At the Pump Room Gardens of Royal Leamington Spa (pages 14, 50), the Lime Tree Walk skirted two sides of an open area and was elegantly adorned with eight iron arches and gas globe lights in 1875. Perhaps the authorities were responding to Granville's criticism of his 1840 visit, when he praised the considerable and varied opportunities for walking and exercise at the competitor spa of Cheltenham which demonstrated 'a superiority over Lexington [Leamington] which the latter can never attain'.

In 2012 this avenue and its remarkable historic globe lights were restored to mark the Diamond Jubilee of Queen Elizabeth II. As a joint project of the Leamington Society, the Central Leamington Residents Association, the Royal Leamington Buildings Conservation Trust and the Warwickshire Gardens Trust this major restoration proves the degree of interest in maintaining the heritage of this spa landscape.

The Linden Arches restored on the walk with new lime trees and globe lights.

Spa Parks

This section contains a selection from the over 60 British spa parks identified (page 114).

1703 The Grove, Tunbridge Wells

1776 New Vauxhall Gardens, Bristol Hotwells

1700s Orange Grove/Harrisons Walks/Parade Gardens, Bath

1794 Sydney Gardens, Bath

1814 Gloucester Park and Gloucester Blenheim (Vauxhall)

1817 Montpellier Gardens, Cheltenham

1825 Calverley Park Grounds, Tunbridge Wells

1824 Queen's Park (The German Spa), Brighton

1828 Manor Baths Park, Askern, nr Doncaster

1830 Pittville Park, Cheltenham

1830 Victoria Park, Bath

1831 The Lawns, Beulah Spa, South Norwood Royal

1834 Jephson Gardens, Leamington Spa

1834 Dorton Spa Gardens, Buckinghamshire

1860s Promenade Gardens, Malvern

1871 Pavilion Gardens, Buxton

1873 Rock Park, Llandrindod Wells, Wales

1881 Pavilion Spa Garden, Strathpeffer

1882 Hove St Ann's Well Gardens, Brighton

1885 Pavilion Gardens, Matlock Bath

Tunbridge Wells
The Grove 1703

At 'The Wells', discovered on a featureless heath in Kent by Lord North in 1606, there exists a very early public green space near the visitor lodgings at Mount Sion. As early as 1703, the Earl of Buckingham presented this woodland to the town along with £800 to be invested for the maintenance of a park under the stewardship of trustees. Walks were created to allow 'bibbers' (water drinkers) to wander, in their social groups as part of their daily routine at the spa and it was placed in trust 'to be continually preserved for a grove and shade, and walks, for the use of all the inhabitants'.[90]

Situated 'within a short distance of the Chalybeate Springs, Parade and Libraries, with paved causeways thereto' the park was a key amenity at this spa.[91] Now it is a public park with prominent signage to explain the park's history.

Vintage early 20C postcard of The Grove, Tunbridge Wells.

Bristol Hotwells
New Vauxhall Gardens c.1776

The stunning setting of the Avon Gorge and the riverside location of the 'Hotwells' were a major attraction. Described in 1799:

> *'The Downs near to the Wells are covered with verdure all the year, and abound with odoriferous plants and herbs: they are enlivened by agreeable prospects on every side...'*[92]

Modern view of the 'verdure' in the Avon Gorge, from the location of the former Rownham Ferry service to the New Vauxhall Gardens.

These prospects lent themselves to offer excursions via the Rownham Ferry (Image page 67) to the opposite shore, where the 'New Vauxhall' pleasure gardens were created to attract the spa visitors. These grounds were abandoned as spa visits declined and any vestiges were cut through by a new railway line in the 1840s. Until the 1980s Rownham Hill Wood was a horticultural teaching area for a local college, with fruit trees, cultivated raspberries and mint growing among woodland plants. Now the site has been replanted as a woodland garden with emphasis on ecology. There is signage to describe the nature conservation project here, but just as there is no remaining physical evidence of the New Vauxhall gardens, neither is there mention of this former spa park.

Bath
Orange Grove/Harrisons Walks/Parade Gardens 18C

From at least the 1720s one of the earliest places for perambulation close to the springs became known as 'The Grove', renamed Orange Grove in honour of the Prince of Orange who visited in 1734. It consisted of a formal lay-out of trees, gravel floor and a pavement in front of a row of shops.

> *This rendezvous of fashion ... seven parallel rows of lime-trees spread their shade en quinconce over the open area.*[93]

Public promenading had already been well-established in Bath by the time an 1819 guidebook sang its praises, describing 'a charming open area, planted with rows of elm trees...

> *Whether thou art wont to rove*
> *By PARADE, or Orange Grove,*
> *Or to breathe a purer air*
> *In the CIRCUS or the SQUARE:*
> *wheresoever be thy path*
> *Tell, O tell, the joys of BATH.'*[94]

By 1840 it had changed trees again; A. B. Granville mentioning 'The present circle of sycamores, planted of late years around the obelisk'. It became reduced to a railed garden and now is no more than a traffic island behind Bath Abbey.

Orange Grove was superseded by walks along the River Avon, and a triangular space overlooked by 'The Parades' offered a central bowling green. This riverside park survives as Bath's much photographed showpiece 'Parade Gardens', which now presents as a typical Victorian public park complete with lawns and island bedding and a bandstand. There is however still a fee payable for visitors to access the park, making this a true resort park in the spirit of former subscription gardens at spas.

View of Bath from North Parade onto Parade Gardens. J. C. Nattes, 1806.

Parade Gardens, Bath, with the historic dome-capped kiosk at the top of the steps for collection of entrance fees. Craig Roberts / Alamy

Bath
Sydney Gardens 1794

This park is regarded as one of Britain's best remaining Georgian 'Vauxhall' pleasure gardens. Its use as a spa park is not often mentioned, but the proprietors of Sydney Gardens promised in 1830 not just 'openable gardens for morning promenades', but also spa waters from Bath *and* other spas 'for the use of the company'.[95] As a favourite for social intercourse and grand outdoor events for the spa clientele, it is particularly highlighted as having been a favourite place for Jane Austen who was resident in Bath for some time.

The gardens lie on sloping ground and were created following previous pleasure garden ventures across the opposite bank of the river from the abbey and spa facilities. After a few decades, it was cut through by both a canal (1810) and then a railway line (1840), with bridges linking the opposite sides of the park. This actually led to a further fascination with Sydney Gardens for viewing these innovations at Bath.

As an important heritage landscape, it was awarded a National Lottery Heritage Fund grant in 2019 for its restoration.

The Kennet & Avon Canal cutting through Sydney Gardens.

Gloucester
Gloucester Park (Rigney Stile Spa Pleasure Grounds), 1820s

Well south of the Cathedral at Gloucester, we have an example of a rather short-lived spa discovered in 1814 which left a legacy of a fine public park in a residential suburb of the town (image page 13). In 1840 it consisted of two green paddocks 'skilfully laid out in shrubberies and wide gravel walks'.[96] A. B. Granville was impressed with the Rigney Style Grounds with walks 'shaded by belts of trees to one side, which impart to the whole the character of a private park.'

Tree-lined gently curving walk for appropriate exercise at Rigney Stile Spa.

Now renamed Gloucester Park, it is still a local recreational park with sport facilities for cricket and bowling and additional public park features such as a playground. Information boards provide excellent detail on the spa-based heritage of this park.

Gloucester Blenheim (Vauxhall) Garden 1814

As at Bristol, associated pleasure gardens were created. This 'Vauxhall' was a commercial enterprise directly linked to the spa nearby. When the spa declined - inevitable due to competition from the superior destinations of Cheltenham and Bath not far distant - these gardens were lost in the residential streets built in the 1840s and only the 'Vauxhall' Public House remains, whose carpark may have been the former bowling green.

Cheltenham
Montpellier Gardens 1817 (Image page 36)

Described by A. B. Granville as the 'gay, varied and improving gardens and pleasure grounds of the Montpellier [spa]', this early park featured trees and shrub planting, including exotic specimens as a backdrop to its varied entertainments on offer to spa visitors. Another Well Walk leading up to it was part of the subscription for this garden.

Tunbridge Wells
Calverley Park Grounds 1825 (Image page 83)

Well after the 'hey-day' of Tunbridge Wells' popularity as a spa, the architect Decimus Burton designed a prestigious housing and hotel complex, with associated landscaping about one mile north of the chalybeate well. There were lawns and beds of roses, herbs and shrubs laid out in the grounds.[97] Despite being some distance from the town, it served as an attraction for a fine clientèle. This park was actually the space that the garden historian Christopher Thacker (page 12) had in mind as the early public park in Tunbridge Wells.

THE BAND PAVILION, CALVERLEY GROUNDS, TUNBRIDGE WELLS
5547

Brighton
Queen's Park (The German Spa) 1824

The fashionable seabathing resort of Brighton capitalised on Dr Frederick Struve's enterprise in concocting 'artificial' waters similar to those offered in continental spas, so becoming known as the 'Royal German Spa'.

The park was originally laid out as a 'subscription pleasure garden' around an imposing pump room as part of a plan to develop an exclusive estate of villas around the park. Though not completed, two Charles Barry designed villas and the two formal entrance arches remain. A building with rebuilt pump room façades houses the Royal Spa Nursery School.

The rebuilt façade of the former pump room at The German Spa, Brighton.

Regencysociety.org

Askern, nr Doncaster
Manor Baths Park c.1828

This town is situated adjacent to a large lake which is fed by several natural springs rich in minerals. In 1786 a first bathing house was built, then in 1828 rebuilt as the Manor Baths with landscaping along the margin of the lake including a bowling green and walks winding through luxuriant plantations.

Further baths and the Askern Spa Hydro hotel opened in and near the town in the nineteenth century, but all declined after a colliery opened in 1911 and Askern became a coal mining town. In 1924, the Spa Hydro actually became the Miners Welfare. The town, however, did continue to benefit from the fine scenic spa park, which retained its tree-lined walk by the lake with boating, fishing, bathing, bowling green and tennis in what became a very industrialised, urbanised area.[98]

Cheltenham
Pittville Park, 1830

This was the park that came top of Christopher Thacker's list of early parks, laid out around the grand John Forbes-designed Pittville Pump Room at the northern edge of town (page 70). Developer Joseph Pitt's residential property scheme was intended to create a new spa town of up to 600 houses over 200 acres; in the event the project was never fully realised. But integral to Pitt's plans was the creation of an elaborate 82 acre spa park with lawns, lake, elegant stone bridges, rock garden, ornamental trees and luxuriant shrubberies laid out by landscape gardener Richard Ware. [99] As with many of the spa parks, it was private to the residents and subscribers of the spa, only becoming a public park in 1894.

Joseph Pitt's architect Robert Stokes (1809-80) designed these gate into the spa park. *Photo © Stephen McKay (cc-by-sa/2.0)*

Bath
Victoria Park 1830

The early public park in Bath identified by Thacker was one mile from the Pump Room, but designed for spa visitors, nevertheless. Described as 'A Royal Kurgarten'.[100] the creation of this park was thoroughly linked to Bath's status as a spa resort. An 1829 council meeting found that Bath, 'so blessed by Providence for the acknowledged virtue of her healing springs', should address its shortcomings in terms of 'shady promenades and agreeable drives' by emulating Continental and other British resorts where

> *'a pedestrian may enjoy exercise shaded from the noon day sun, or the invalid be invited to an airing without the dust or inconvenience of a turnpike road.'*[101]

Princess Victoria made her appearance at the 'auspicious Opening of the Park' on Saturday, 23rd October 1830. Named in her honour, the Royal Victoria Park was one mile distant from the source spa waters and therefore was more accessible to those who had the finest lodgings at the Palladian Royal Crescent than for use as a promenade for those taking the waters at the Pump Rooms. But it was in the vanguard of public park development exactly because its cost could be justified by the advantages it would bring to the town as a fashionable visitor destination. The then city architect Edward Davies incorporated walks, rides, picturesque paths, lakes and an arboretum, providing diversion and indeed the 'rational recreation' that became a theme for the Victorian public parks, 'intended to instruct and delight'.[102]

Nowadays it remains as a fine public park with an imposing entrance, tree-lined walks, open space, pond and play area in the best tradition of what we have come to expect of our city parks.

Beulah Spa, South Norwood
The Lawns 1831

A chalybeate well in south-east London spawned a most fashionable spa and gardens, created by an illustrious architect at the time: Decimus Burton. 1850s illustrations show undulating lawns, winding paths, clumped trees and a formal round bed with a tree on a raised mound in J.C. Loudon's Gardenesque style that was popular at that time. Climbers were trained over the extraordinary rustic wigwam pump house, and a rose garden, lake, maze and 'orchestra' were part of the grand design.

The 25 acre pleasure garden hosted spectacular events, including a visit from Johann Strauss Snr conducting his orchestra with 10.000 visitors listening and waltzing in the park. Its location just two miles south of from Sydenham Hill meant that Beulah's popularity was short-lived, as the superior attraction of the relocated Crystal Palace in the mid-1850s reduced its custom tremendously. Surviving are woodlands with the Beulah Spa Oak, and a Linden Tree planted by Kaiser Wilhelm, now standing in the suburbs.[103]

Royal Leamington Spa
Jephson Gardens (Ranelagh Pleasure Gardens), 1834

This park is situated opposite the Leamington Pumprooms which has its own adjacent spa park (pages 14, 50). Originally this park was called 'The Ranelagh Pleasure Gardens...tastefully laid out and furnished'. By 1834, it was enlarged and in 1846 it was further embellished and renamed in honour of a Dr Henry Jephson who had helped to promote the healing properties of the town's spa waters and build houses for the town's poor residents. He personally recommended exercise for spa patients, especially promenading, and in these gardens free entry was permitted 7-10am, after which they were open to paying customers and patrons of a nearby spa bath house.[104]

The gardens have, over the years, enlarged and changed hands and now they present as a glorious public park in the best tradition of a 'Kurpark'.

The Jephson Gardens, Leamington c. 1910 - in Lenton & Sons 'Grand Photographic View Album of Leamington Spa & Shakespeare Country'.

Buckinghamshire
Dorton Spa Gardens, 1834

This now lost spa landscape is one of the minor nineteenth century spa ventures attempting to create facilities for a fine health resort. This spa was located in a rural, remote location where a pump room was built after promising cures were described in an 1833 treatise on its chalybeate waters. Included in this document was the instruction that the garden layout was to 'add to these buildings all the agremens of a pleasure-ground.'[105]

Dorton pump room and bath house with newly laid out gardens in the 1830s.

The intended 1834 plans were never fully realised, but a truncated version of the classical spa building stood in a woodland clearing of twelve acres of garden with terraced planting of evergreen and deciduous shrubs, curved paths and beds around a serpentine lake.[106] It is clear from an 'act to grant further powers of leasing' that new owners were keen to further develop land here for large-scale building and landscaping 'by reason of increasing resort and the want of sufficient accommodation'.[107] However, without the availability of suitable accommodation from the start and practically no entertainment opportunities for visitors to this exceedingly rural spot, the spa did not thrive and within 15 years had declined, sadly left to go to ruin and nature reclaiming the manicured gardens.

The decaying spa building was demolished c.1906 and survives only as footings, while the gardens are lost and remain only as hollows, banks and 'the generally dry remains of a boating lake and some mature yew'.

Remains of Dorton Spa shortly before demolition in the early 1900s.

We are lucky to be able to piece together the history of lost landscapes such as these thanks to the efforts of the local Garden Trusts, whose research projects collate and record designed parks and gardens in their region. At Dorton, the heritage of this former spa park was unravelled by one of the volunteers belonging to the Buckinghamshire Garden Trust.[108]

Malvern
Promenade Gardens, 1860s

These gardens were developed during the early excitement of Malvern as the first location for hydropathic baths. The 'water cure' was developed at Graefenburg in Austria by Vincent Priessnitz and under the influence of the Doctors James Wilson and James Manby Gully it was promoted in England from 1842.

Exercise was taken as part of the medicine and there were numerous walks and park areas that were available at Great Malvern. St Ann's Well, the historic water source with the celebrated zigzag walk leading up to it (page 73), had a thoroughly Victorian park created near the base of this walk: the 'Great Malvern Promenade and Nursery Gardens'.

One of many vintage postcards available of the Promenade Gardens, Malvern.

This was an early venture for Birmingham nurseryman and landscape gardener Richard H. Vertegans who was later commissioned for other prominent public parks in Wolverhampton and Handsworth. He, in effect, created a commercial spa park to tempt the health visitors to walk, enjoy the view, hear music and purchase plants from his nursery enterprise. From at least 1860, the gardens held an annual flower show and a German band stationed at Malvern played in the bandstand[109].

GREAT MALVERN PROMENADE AND NURSERY GARDENS.

THESE beautiful Gardens having undergone EXTENSIVE ALTERATIONS, are now OPEN to the PUBLIC.

Proprietor:
R. H. VERTEGANS,
769] Chad Valley Nurseries, Edgbaston, Birmingham.

An upgrade to the Promenade Gardens was promoted in the local *Hereford Times* in July 1867 and it is clear from a later (1869) advertisement that the gardens had an entrance charge. For their 6d, ticketholders had access to a park with lawns and wide pathways, benches, manicured beds including box-edged rose beds, urns, fountain and the bandstand musical entertainment.[110]

MALVERN PROMENADE GARDENS,

Near the ABBEY CHURCH.

These delightful GARDENS are now OPEN. Admission 6d.
Each Visitor is entitled to a Bouquet of Flowers.
The BAND will Play from Three to Five o'clock on Mondays, Wednesdays, and Fridays.

425 R. H. VERTEGANS, Proprietor.

The garden lay-out, ticket arrangements, music offer and indeed its hill-side location was very much in the style of European spa gardens, possibly influenced by the connection with the hydropathic movement. The commercial spa park at Malvern changed hands but continued to operate with bands playing into the early twentieth century. Nowadays the former spa garden is part of the private Rose Bank Gardens, with the upper 'promenade' still clearly visible from the road.

Buxton
Pavilion Gardens, 1871

At Buxton Pavilion, a twelve acre spa park was laid out in the hey-day of Public Park development. The accomplished landscape architect, Edward Milner, who had worked with Joseph Paxton who designed the Crystal Palace, was commissioned for the design to ensure continuing appeal for this northern spa.[111] Considered one of the best examples of the 'Gardenesque' style, the park displays a combination of formal beds set against a heavily planted background and individual trees and shrubs laid out to exhibit their full form as advocated by the gardening encyclopaedia publisher J.C. Loudon. Winding walks, ornate bridges, bandstands, and a lake completed the scene. Description from an 1886 guidebook enthuses,

> *Extensive acreage of gardens and pleasure-grounds .... artistically laid out, embellished with all the aids of modern gardening...The walks...not less than two miles in length...and a succession of floral beauty at different seasons.*[112]

Vintage Postcard of Buxton Pavilion Gardens featuring leafy walks.

Llandrindod Wells, Powys, Wales
Rock Park, 1873

In the 1840s this rural Welsh town had only 270 inhabitants.[113] Even now, its population is little more than 3000, yet it benefits from a fine public park laid out after the coming of the railway in the 1860s gave the town's spa industry a boost. Indeed, Rock Park is one of the earliest public parks in Wales with a network of paths through native woodland and a Victorian arboretum. A viewpoint from a natural rock outcrop to the river provides picturesque interest and there are extensive outdoor bowling greens.

Detail from Vintage postcard: Rock Park and Pavilion.

Even into the twentieth century the spa continued to function (page 27) and a band was employed during the season to play in the open pavilion. An advertisement from 1997 for the 'Rock Park Spa' proclaims its setting in the 'beautiful, tranquil surroundings of Rock Park'.[114]

The park area, the original pump rooms and bath house have been restored in 2008 under a Tourism Capital Scheme managed by the Spa Town Trust.

Strathpeffer
Pavilion Spa Garden 1881

The mineral springs of this Scottish village were under the management of the Cromartie Estate in the mid-eighteenth century when the Countess of Cromartie (Duchess of Sutherland) became keen on promoting the village as a spa location. Familiar with the fashions and facilities at Continental spas, the Countess was instrumental in the lay-out of the village and adaptation of the grounds next to a new pump house. The surrounding park provided level greens for bowls, tennis and croquet as well as promenades among flower beds. A network of wide footpaths paths was constructed in 1897 with suitably shallow steps on the sloping ground of the adjacent woodland.[115]

In the twentieth century, the pavilion's deteriorating condition and lack of profitability resulted in a lack of interest in maintaining the building and grounds. This changed in the mid-1990s when the spa grounds and surviving Upper Pump Room were sold to the Highland Council for a nominal sum and now are maintained for sport, leisure and events. Even now, the Strathpeffer golf course is a key attraction of this Highland resort.

Strathpeffer Pavilion Spa gardens remain acknowledged by Historic Environment Scotland as a specialised landscape within the resort which proudly terms itself 'Victorian Spa Village'.

Brighton
Hove St Ann's Well Gardens 1882

The chalybeate well one mile from Brighton had a setting considered to be a most picturesque spot, commanding a beautiful view of meadows down to the sea. The park developed from the 1800s with a pump room built in 1882. Copious serpentine paths leading through mixed trees were described at the time as 'six acres of refreshing foliage and shady walks', providing destinations such as a pond, picnic field, bandstand; even a hermit's cave in the Picturesque tradition (page 42).

The development of these gardens during late Victorian times produced a combination of spa park, pleasure garden and public park, with admission being charged for a wide range of amusements including fairground-type attractions such as rifle ranges and boat swings.[116]

Victorian photo displayed on an information board near the location of the former pump house at St Ann's Well Gardens at Hove, Brighton.

The rich private owners (Goldsmid family) refused to sell their park for building development and let the Council purchase it at a reasonable sum for use as a public park, opened in 1908. The intersecting walks still wind among many mature trees on this attractive hillside and although the pump house was demolished in 1935, the spa heritage of the park is acknowledged via information boards throughout the park.

Matlock Bath
Pavilion Gardens, c. 1885

The original wooded area south of Matlock Bath Pavilion (Kursaal) had a number of thermal springs with their outfall to the river. The area had been cleared of trees and was used as a rubbish tip for a while until the 1880s when sixteen acres of newly landscaped gardens 'unrivalled for their picturesque beauty', became part of a scheme that included a roller coaster and other amusements.[117] Matlock's reputation was by that time more as a pleasure resort rather than a spa, as taking the waters was definitely no longer a major activity in Britain, but travel for personal and cultural leisure was. A number of landscaped areas were amalgamated in the 20th Century, becoming the now public Derwent Gardens.

Matlock Bath Pavilion and Gardens from 'Matlock Visiting List, Oct 1885.

Private Spa Gardens

These encompass the landscaped gardens in private establishments offering spa waters or hydopathic treatments. The gardens therefore tended to cater principally for their own guests, often resident. As such, they were less obviously designed for the cure itself and more as an attractive space for visitors, their status as spa-specific landscape is more ambiguous.

A small selection of interesting examples is given.

1708 Beacon Hotel Cold Bath and Tea Garden,
Rusthall, Tunbridge Wells

1700s Spa Gardens, Islington, London

1820s Winkfield Spa Pump Room Gardens, Windsor

1834 The Crown Hotel Baths Gardens, Harrogate

1830s Harlow Carr Hydro Hotel Garden, nr. Harrogate

c1840 Dol-y-Coed Hotel Gardens Llanwrtyd Wells, Powys

1837 Victoria Spa Gardens, Bishopton, Warwickshire

1838 Grounds of 'The Spa' Hotel, Tunbridge Wells

1893 Brine Baths Hotel garden, Nantwich

Rusthall, Tunbridge Wells
Beacon Hotel Cold Bath and Tea Garden 1708

In the vicinity of Tunbridge Wells, on a ridge past 'The Common' an establishment offered an ornamental garden with a plunge pool, providing a spa bathing opportunity out of town. Cold bathing outdoors was (understandably!) superseded by hotel-based bath facilities, and only the foundations of its structure now survive, along with the spring and remnants of an avenue of mature yew trees (now in poor condition).

View towards Tunbridge Wells across the lawns of the Beacon Hotel.

In 1818, some new tea gardens were separately laid out on an adjacent ridge, and these became an established part of the local visitor attractions. Even now, the plunge pool area remains as a wooded valley adjacent to the 1891 Beacon Hotel which is still approached via 'Tea Garden Lane' [118]

Islington
Spa Gardens c.1700

This an historic example of one of the once numerous spas of the London region which was swallowed up and superseded by development of the urban environment. Many of the eighteenth century London 'Spaws' tended to be day-visitor attractions of the 'tea-garden' variety and once the building of streets and houses became of greater importance than a green environment, spa gardens were easily given up for development. Pleasure grounds such as Islington's 'New Tunbridge Wells' with a reputation sufficient to attract visitors such as John Evelyn and Lady Wortley Montagu in the 1680s were completely lost by the Victorian era and now only names and a few wells remain.[119]

The Islington Spa of 1684 was in fact created in an area essentially rural in character, the garden providing a 'man-made' green space laid out around a central mineral spring basin, surrounded by a balustrade with shading lime-trees and pretty arbours. A contemporaneous description mentions 'pedestals and vases grouped with taste under some extremely picturesque trees'.[120]

Detail of Islington Spa, 1733 engraving by George Bickham Jr., based on a May Day Illustration 1720.

Even in 1788 the spa was still hailed as a 'charming spot of beautiful rusticity'. However, when trade reduced and the coffee house was demolished in 1810, the gardens were reduced in size and two rows of houses called Spa Cottages were erected in the garden in the 1840s. The spring ceased to flow by 1860, abstraction of water for a growing population had this effect at most of the London spas. Streets and houses were built all around, although there is mention in the 1890s of a little public garden of 'Spa Green'. The site of the spring was described as 'preserved in an obscure nook, amidst poverty-stricken and squallid (*sic*) rookery of misery and vice'. In the early twentieth century this overcrowded area was designated for slum clearance and World War II bombing took its toll.

A brand new regeneration project by the modernist architect Berthold Lubetkin arose there after the war 1946-49, creating 'Spa Green Estate' with the residential 'Wells House' at its centre, a significant development as witnessed by its Grade II* listing in 1998. Thus, the lost spa garden at Islington lives on in name and the principle of green landscaping to enhance the environment was revived.

Part of the modern Spa Green Estate. *Photo © Jim Osley (cc-by-sa/2.0)*

Windsor
Winkfield Spa Pump Room Gardens 1820s

J.C. Loudon's *Gardener's Magazine* of October 1829 reported that in Windsor Forest, 'a spring has been discovered...superior to Cheltenham Water'. The erection of a rustic pump room and the use of a nearby mansion as hotel and boarding house is described. Crucially, the article mentions the creation of extensive walks 'in a country affording abundant food for the botanist and the lover of the picturesque'.

The clear intention to create a suitable outdoor environment for health seekers makes this a spa garden for what appeared to be a very short-lived, speculative project. The 1829 illustration and description give the layout with a circular lawn fronting the building, surrounded by walks in 'natural coppice wood'.[121]

Illustration of a proposed new spa in Windsor Forest . Gardeners Magazine, 1829.

Many such private spas were planned or initiated, but most did not get off the ground commercially. Certainly, we hear no more of this spa, whose former location is now within the Windsor Legoland site.

Harrogate
The Crown Hotel Baths Gardens 1834

In 1626 the Crown Hotel's neighbouring Old Sulphur Well was the subject of a famous book. The happy discovery of a sulphur well actually within the grounds of the hotel in 1822 prompted a Mr. Thackwray to lay out the 'Montpellier Gardens' at the hotel around 1834 and to erect the original baths therein. Whilst there was controversy over access to the mineral waters leading to a new Act of Parliament in 1841, the Crown and its facilities were so popular with health visitors that it was nick-named 'The Hospital'.

In 1889 the Crown Hotel's gardens were sold to Harrogate Corporation and became subsumed in other development. However, the hotel still operates and has maintained its floral tradition in typical spa embellishment style as well as capitalising on its spa heritage with historical information and images (such as the twentieth century photo below) throughout the hotel.

1960s Photograph displayed inside Harrogate Crown Hotel bedrooms, along with historical information about its spa history.

Harrogate
Harlow Carr Hydro Hotel Garden 1830s

A familiar name to gardeners, the current RHS Gardens at Harlow Carr incorporate the former spa garden of the Hydro Hotel, almost two miles from the Crown Baths at Harrogate.

Engraving showing the Harrogate Arms, Bath House and gardens, c1857.

The mineral well in the foreground of the image above, was sited in the grounds with a network of wide pathways and planting in the Gardenesque style of the age.

Llanwrtyd Wells, Powys, Wales
Dol-y-Coed Hotel Gardens C18

A spring discovered in 1732 prompted the conversion of a private house to a thriving spa hotel. The grounds led to a lively small river the wells closed circa 1960. The building no longer functions as a hotel and is now used as offices.

Bishopton, Warwickshire
Victoria Spa Gardens, 1837

Near Shakespeare's Stratford upon Avon, a chalybeate well offered another opportunity for a spa enterprise. A consortium of local businesspeople were keen to develop a fashionable inland resort, building a hotel which was blessed by the patronage of a visit by Queen Victoria herself. This was a good start to help offset the fact that a small spa relatively close to fashionable Cheltenham Spa was unlikely to draw many visitors.

Suitable gardens were laid out around the building as villa-type pleasure grounds with walks and a pond. An 1840 description mentions a flower garden, extending about an acre, 'neatly laid out in green plots, with a piece of water in the centre'.

Gardens at the Victoria Spa, Bishopton nr. Stratford upon Avon c. 1840.

By the 1870s the spa and associated development failed, said to be due to lack of local entertainment. The Victoria Spa Lodge and Pump House remain in the private grounds that have inevitably altered since spa times. The Lodge is now a guest house that also has a historic connection to Bruce Bairnsfather, WWI cartoonist of 'Old Bill' fame.

Tunbridge Wells
Grounds of 'The Spa' Hotel, 1878

North of springs at Tunbridge Wells, the elevated area overlooking the common had provided elegant lodgings for centuries in an area named 'Mount Ephraim'. Many of these large houses converted to hotels and guest houses, including this impressive 'Spa Hotel'. A full range of hydropathy (bathing) treatments were offered at this establishment, rejuvenating the town's fading reputation as a spa for taking the waters. Emphasised are the 'elevated yet sheltered position ... overlooking the far-famed Common and an extensive prospect of fine country...the grounds are 53 acres in extent and embrace the most charming and distant views'.[122]

A large Edwardian painting displayed inside Tunbridge Wells Spa Hotel.

Nantwich
Brine Baths Hotel garden, 1893

This is just one of numerous Hydro Hotel gardens in estates with existing parkland or garden which were enhanced for serving spa visitors. Here a brine spring prompted investment and conversion of 120 acres of parkland for use by hotel guests.

Detail 1930s postcard Brine Baths Hotel, Nantwich.

The house - originally Shrewbridge Hall - was converted into an hotel with an impressive carriage sweep leading up to it and fine gardens including an ornamental rose garden as well as tennis and croquet courts.[123]The hotel closed in 1943, the building demolished 1959 and the entire site developed for housing in the early 1970s.

PART III
DATA ON BRITISH SPA LANDSCAPES

Analysis of Spa Landscaping

The history of spas in Britain was studied to identify those for which evidence could be found of landscaping deliberately associated with its spa function. Particularly useful for research was the study of historic maps, aerial views, paintings, photographs and other images. Historic descriptions and correspondence as well as on-site visits also contributed to the material presented.

A. B. Granville's 1840s volumes on English spas provided occasional comments on planting and landscape, as did other contemporary writers. Much information was gleaned from the entries and drawings of Kathleen Denbigh's *A Hundred British Spas* written in 1981.

Schulze's categories of recognisable elements within a suite of designed landscape at spas (page 39) served as a starting point for collection of quantitative data. The figures gleaned from this study are, however at best preliminary, as information is widely dispersed geographically and over time and the ephemeral nature of gardens means that many examples are likely to have been missed.

Of the 345 spas, springs and watering places in existence at some time in Britain, around 100 spa locations were of sufficient size and importance to be mentioned in compilations of British spas. These were the ones analysed to establish the existence and nature of any deliberate spa landscaping. Of these, 62 spas had at least one or more of the elements typical of spas, although 16 were wells developed in already existing parks or gardens.

It is notable that there were over 50 spas for which no record of planting specifically to serve or enhance the spa facilities was found. Often the apparently unlandscaped spas were just in open country without planted embellishments, others were located inside urban buildings. These have been excluded from the gazetteer of spa landscape further on in this section.

Spa avenues	Park landscaped as spa park		
Well Walk, (A) and promenades Close to well	Proximal Park, (P_p) Close to well	Distal Park, (P_d) At distance from well	Existing Park, (P_e) Well used in pre-existing park/garden
33	43	10	16

At least 43 spas had a spa park close to the wells or pump room, and almost a third of the approximately 100 spas of any consequence had an identifiable well walk or spa promenade. Of course, larger spas invariably had both of these. A few of the earlier spas also had parks away from the water source laid out as pleasure gardens as part of their resort offering. On a smaller scale there were tea gardens attached to hotels providing further diversions for spa visitors.

If we break down the list of spas into categories of their relative size and importance, we find that without exception the early national spa towns such as Buxton, Tunbridge Wells and Cheltenham all provided well walks and spa parks for the benefit of their clientèle. Local spas of repute that developed later would emulate these towns hoping for similar patronage and promenades and spa parks also became part of their offering.

The decline of the spa 'cure' in Britain inevitably resulted in altered land use at former spas where pressures for housing, industry or transport routes were an issue. Of the just over 100 spa landscape elements identified in Britain, (comprising all well walks, avenues and spa parks) only 58 remain in any form and at least 30 of them have become public parks. Of the 43% lost spa landscapes, some may have been very short-lived, reflecting the fortunes of the associated spa.

Landscape elements	Lost (L)	Vestiges or extant trees (V)	Now Private Gardens (G)	Now Municipal Parks (M)
Avenues/ Well walks (33)	15 (45%)	18 (55%)		
Parks & Gardens (69)	29 (42%)	1 (1%)	8 (12%)	31 (45%)
All types (102)	44 (43%)	19 (19%)	8 (8%)	31 (30%)

<u>UNDERLINED</u> spas recorded in a heavy bordered frame entry in the Gazetteer are those that in 1895 were still considered as active spas according to a report of the Royal Medical and Chirurgical Society on the 'Baths and Health Resorts of the United Kingdom' in the *British Medical Journal* of August that year.

The century in which the spa commenced operating is given as well as the patronage of selected important historical personalities, which provides an indication of the relative status of the spa at that time. Individual parks and avenues are given as bold entries, with codes for their features as described below.

Gazetteer of Landscaped Spas

The spa landscape categories have been organised as follows:

- The Well Walk and other tree-lined avenues for promenading have been grouped together as 'Avenues' (A).
- Other path systems have become part of the planting area described here as 'Park'.
- Parks have been categorised according to location, where a 'proximal' park (Pp) is directly associated with the source of the mineral water and therefore visited as part of the curative regime. A 'distal' park (Pd) is at some distance from the well or baths and will have played a supportive function in terms of entertainment for visitors to spa.
- The use of shrubs and ornamental planting has been noted qualitatively only where evidence exists for a period during which the location was actively in spa use for 'taking the waters'. Built elements and park leisure facilities were not quantified but noted qualitatively in selected examples to illustrate their contribution.

Signpost in Rock Park, Llandrindod Wells, Powys.

ELEMENTS:

A – Avenues or 'Well Walks'.

Pp - Park, proximal. Located at, or in near vicinity of, the mineral spring or well for assembly, waiting, imbibing, walking and resting near the source.

Pe - Existing private park or garden developed as a spa where subsequent planting and design could serve a spa-related purpose.

Pd – Park, distal. Located not in the immediate vicinity of the well. To identify as a spa park, it would be designed principally to appeal to spa visitors.

CURRENT STATUS:

L - Lost. No remnants of planting created in association with the spa function remain.

V - Vestigial. Visible planting elements or landscape design remain, incorporated into landscape or urban development.

G - The landscape is at least partially in existence but as a private park or garden.

M - Municipal. The landscape has been repurposed for use as public park or recreational space.

MH – Heritage. The landscape is operated as a public park but additionally its history as part of a former spa is recognised and communicated by means such as information boards, trails, leaflets or guided walk.

ENGLAND OUTSIDE OF LONDON

ADMASTON (WELLINGTON) SPA, Shropshire C18
Pump Room Gardens

ASHBY-DE-LA-ZOUCH, Leicestershire C19
Ivanhoe Baths Gardens (Hall Meadow) Pp, MH, 1828

ASKERN, South Yorkshire C18
Manor Baths Park Pp, M

BATH, Somerset Roman and from C16
1613 Anne of Denmark (James I's queen)
1663 Queen Catherine of Braganza
1702 Queen Anne
1734 Prince of Orange,
1781 Admiral Lord Nelson,
1801 Jane Austen
1830 Princess Victoria
The Grove (Orange Grove), Pp, L c. 1730
Harrison's Walks (Parade Gardens) 1709, Pp, MH
Spring Gardens (Pleasure Garden) 1740s-1790s, Pd, L
Sydney Gardens Vauxhall, Pd, MH. 1796 Charles Harcourt Masters.
Royal Victoria Park, Pd, MH. 1830 Edward Davis

BRISTOL HOTWELLS, Avon C17
1677 Queen Catherine of Braganza
1810s Queen Charlotte
Hotwells Old Pump Room Riverside Walk, A, L
New Vauxhall Gardens, Pd, L

BISHOPTON (Stratford-on-Avon), Warwickshire C19
1838 Queen Victoria
is Victoria Spa Lodge' and Pump House Garden, Pp, G

BRIGHTON, East Sussex C18
1830 Mrs Fitzherbert, Mistress to George IV
1830s William IV & Queen Adelaide
German Spa (Queen's Park) 1824
Hove St Ann's Well Gardens 1882, Pe, MH

BUXTON, Derbyshire C16
1570s Mary Queen of Scots
1697 Celia Fiennes
The Walks 1820s Pp, MH
Pavilion Gardens Pp, MH. 1871 by Edward Milner/ Joseph Paxton

CANTERBURY C17
1697 Celia Fiennes
Mr. Masters' Exotic Nursery' 1830s Pe, L.

CHELTENHAM, Gloucestershire C18
1788 King George III
Well Walk A, L, 1740 Henry Skillicorne,
Sherbourne Spa (Imperial) Gardens Pp, M, 1818 G.A. Underwood
Sherbourne Spa Promenade, A, L 1837
Montpellier Wells Walk 1809 Henry Thompson
Pleasure Gardens 1817 *With early Bandstand* 1864
Pittville Park A, Pp, MH, 1830 John Forbes, Richard Ware (nursery)

CHESHAM, Buckinghamshire C19
Mineral Cottage garden, Pp, L. 1820

CROFT-ON-TEES, North Yorkshire C17
New Well Pump Room Gardens, Pp, L

DINSDALE-ON-TEES, Durham C18
Spa Bath Grounds, 1881, Pp, L

DORTON, Buckinghamshire C18
Dorton Spa Gardens 1834 Pp, L designed by *James Hakewill.*

DROITWICH, Worcestershire C19
Brine Baths Park 1895 Pp, M

EPSOM, Surrey C17
1684 Charles II
1690s George of Denmark,
Epsom Common promenade, A, L

GLOUCESTER, Gloucestershire C18
Old Spa House Garden, Pe, L
Rigney Stile Spa Grounds 1815, A, Pp, MH
Blenheim (Vauxhall) Gardens, c.1815, Pd, L.

HASTINGS, East Sussex C19
1882 Edward, Prince of Wales and Princess Alexandra
Alexandra Park, Pp, MH. 1880 Robert Marnock

HARROGATE North Yorkshire C16
1697 Celia Fiennes
1806 Lord Byron
Montpellier Sulphur Wells/Crown Hotel Gardens 1834 Pp, MH
Valley Gardens 1887 Pp, MH.
Harlow Carr (Hydro Hotel) Gardens 1844, Pp, G

HOCKLEY, Essex C19
Pump Room Gardens, Pp, L. Design 1838 James Lockyer

KING'S SUTTON (ASTROP WELLS), Northants C17
1694, Celia Fiennes
Astrop Park, c.1740 A, Pe, V

(ROYAL) LEAMINGTON SPA, Warwickshire C18
1830 Princess Victoria
1838-9 Louis-Napoleon Bonaparte (in exile)
Pump Room Gardens 1814 Pp, MH
Linden Walk 1814 A, VH
Jephson Gardens 1834 Pd, MH
Ranelagh Pleasure Gardens 1811 Pd, L

MALVERN, Worcestershire C18
1831 Princess Victoria,
1843 Dowager Queen Adelaide
1851 Annie, daughter of Charles Darwin
Winter Gardens Promenade A, V. 1885
Hay Well Baths Pleasure Gardens Pp, MH. 1840s
Promenade Gardens Pd, G. 1860s by Richard H Vertegans.
Assembly Room Gardens Pp, MH. 1885
Victorias Walk A, VH 1831

MATLOCK BATH, Derbyshire C18
1794 JMW Turner
1806 Lord Byron
1832, 1844 Princess/Queen Victoria,
1958 John Betjeman
Riverside/Lovers Walk A, V
Pavilion Gardens, Pp, M 1885

NANTWICH, Cheshire	C19
Shrewbridge Hall (Brine Baths Hotel) gardens Pe, V	

NEW MALTON, North Yorkshire	C17
New Malton Gardens, Pe, L c.1810s	

RIPON, North Yorkshire	C19
Pump Room Gardens Pp, A, L. 1834	

SCARBOROUGH, North Yorkshire	C17
1720s Daniel Defoe, Celia Fiennes.	
1732 Sarah, Duchess of Marlborough	
New Spa Hotel Gardens, Pp, M. 1858 Joseph Paxton	

SHOTLEY BRIDGE, Durham County	C19
1839 Charles Dickens	
Spa Drive, A, V 1838	
Spa Grounds Pp, V, 1838	

SOUTHAMPTON, Hampshire	C18
1789 George III	
1755 Horace Walpole	
Quay Walk Promenade A, L	
Spa Gardens Pp, L	

SPITTAL, Merseyside	C19
Spa Well Garden, Pp, MH.	

STARBECK, North Yorkshire	C19
Pump Room Avenue, and gardens A, V	

TENBURY WELLS, Worcestershire C19
Pump Room Gardens, Pp, L. 1862 *James Cranston*

THETFORD, Norfolk C19
1819 William Frederick, 6th Duke of Gloucester
Spring Walk, A, V. 1819

TUNBRIDGE WELLS, Kent C17
1630 Queen Henrietta Maria 1663 Charles II
1698 Queen Anne 1830 Princess Victoria
The Pantiles Lower and Upper Walks A, V , 1638
The Grove Pd, MH. 1703
Beacon Hotel Cold Bath Tea Gardens, Rusthall, Pd, G. 1708 Calverley Hotel Grounds Pd, MH. 1828 Designer Decimus Burton.
The Spa Hotel gardens Pe, G1878

TILBURY, Essex C18
Spa Gardens, Pe, L.

WEYMOUTH, Dorset C18
1794 George III and Queen Charlotte
Radipole Pump Room Gardens, Pp, L. 1831 James Cranston

WINKFIELD SPA, Windsor and Maidenhead C19
Pump Room Gardens 1820s, Pp, L.

WITHAM, Essex C18
Walk Field Avenue A, V

WOODHALL SPA, Lincolnshire C19
Victoria Hotel Gardens Pe, L 1860s

LONDON AND GREATER LONDON

ACTON WELLS	C17
Old Oak Common, Pp, L,	

BAGNIGGE WELLS	C18
Spa tea garden Pp, L.	

BARNET, North London	C17
1664 Samuel Pepys	
1697 Celia Fiennes	
Spa Avenue, A, L	

BEULAH, South Norwood	C19
1820s Michael Faraday (Analysis of waters)	
1838 Johann Strauss Snr (Played concert)	
1830s Charles Dickens	
1830s Queen Victoria	
Beulah Spa Gardens (Spa Hill), Pp, M 1831 Decimus Burton	

BROMLEY	C18
1668 Samuel Pepys	
Bishop's Palace Pleasure Grounds (Queens Gardens) Pe, M.	

FINSBURY/ LONDON SPAW	C17
London Spaw A, Pp, L c.1720	

HAMPSTEAD	C18
Old Well Walk, A, L	

ISLINGTON	C17
1689 Lady Wortley-Montagu,	
1660s John Evelyn	
New Tunbridge Wells Pleasure Grounds, Pp, L	

KILBURN	C18
1772 Oliver Goldsmith	
The Bell Tea Gardens Pp, L.	

POWIS WELL	C18
Lamb's Conduit Fields A, L	

RICHMOND	C17
The Walks A, L.	

SADLER'S WELLS	C17
The Long Walk, A, L	
Theatre Gardens Pp, L. 1680s	

ST CHADS	C18
Pump Room Gardens Pp, L. c.1760	

ST PANCRAS	C17
Old Walk, A	
New Plantation, Pe, L. 1697	

STREATHAM	C17
1670 John Aubrey,	
1760s Dr Johnson	
Streatham Promenades A, L. c.1680s,	
Well House Gardens (Rookery Garden) Pp, M. c.1680	
New Wells House Tea Garden Pp, L	

WALES

BUILTH WELLS, Powys C18
1808 Lady Hester Stanhope
Well Avenue at Park Wells, A, V
Park Wells Gardens Pp, G

LLANDRINDOD WELLS, Powys C18
1926 Prince of Wales (Edward VIII)
Pump House Hotel Gardens 1888 Pp, L
Rock Park Pp, MH

LLANGAMMARCH & LLANWRTYD WELLS, Powys C18
Unusual Barium Spa
Lake Hotel Gardens Pp, G
Irfon Riverside Quay Promenade A, V
Dol-y-Coed Hotel Gardens, Pp, G

SCOTLAND

DUNBLANE, Perthshire C19
Hydro Hotel Lawns Pp, G 1878

MOFFAT, Dumfries and Galloway C17
1752 James Boswell 1787 Robert Burns
Hydro Avenue A, V 1878 Hydro Hotel Gardens Pp, V

STRATHPEFFER, Highland C18
Pavilion Spa Gardens Pp, MH. 1860s

Notes

Preface

[1] Eidloth, 2020.

The Study of Spa Landscapes

[2] Hobhouse, 2002.
[3] Strong, quoted in Conan, 1999.
[4] Layton-Jones, 2014.
[5] Borsay, 2006.
[6] Kaspar, 2016.
[7] https://whc.unesco.org/en/list/1613 [Accessed 25/08/2021].
[8] Christopher Pound, 'The Correlation of Urbanism and Open Space Planning in Health Resorts – C18 Bath as a Prototype Spa' in Eidloth, 2020.
[9] Ian Bradley,' Promenade Concerts, Music Pavilions and Bandstands; The Place of Music in Spa Parks', in Eidloth, 2020.
[10] Anna Jakobsson, 'Ruled by Routine and Ritual: Spatial Organisation of the Spa Environment at Ronneby, South-East Sweden', *Garden History,* Vol. 32, No. 2 (Winter 2004).
[11] Josep Sánchez Ferré, 'Parks and gardens in spa towns. Therapeutic landscape interpretation notes', *III International Symposium on Thermalism and Quality of Life*, (Campus da Auga, Ourense, Spain, 2019).
[12] Dominique Jarrassé, 'Spring Pavilions, Covered Galleries… Architectural Challenges in the Spa Park and Perspectives for a Heritage Endangered in the Long Run', in Eidloth, 2020.
[13] Sue Berry, 'Pleasure Gardens in Georgian and Regency Seaside Resorts: Brighton, 1750-1840', *Garden History*, Vol. 28, No. 2 (Winter, 2000).
[14] Wroth & Wroth and Curl, 2010.
[15] Rabbits, 2018.
[16] Conway & Rabbitts, 2023.
[17] Thacker, 1994.
[18] Robert Slaney 1833, Parliamentary Debates, 3rd Series, XV, p 1,054, debate of 21 Feb 1833.
[19] Johnston, 2015.
[20] Kaspar, 2016.

Spas in Great Britain

[21] Dr Edward Chorley, a physician at Doncaster quoted in the *Northern Spas* section of Granville, 1840.

[22] https://www.thenorthernantiquarian.org/2011/07/24/st-helens-well-thorp-arch/ [Accessed 16/04/2024].

[23] Borsay.

[24] Egan, 1819.

[25] Ibid.

[26] Vasset, 2022.

[27] Granville, 1840.

[28] Denbigh, 1981.

[29] Sophie Vasset, *Murky waters,* 2022.

[30] Curl, 2010.

[31] Granville's 1840 *Spas* certainly included resorts such as Margate, Torquay or Blackpool which were purely for seabathing.

[32] Canterbury Journal, 07 August 1841.

[33] Jan Woudstra, 'Robert Marnock's Retirement Project: Alexandra Park, Hastings, 1876 -82, *Garden History, 51:1 (2023).*

[34] Granville, 1840.

[35] Hembry, 1990.

[36] Egan, 1819.

[37] Buckinghamshire Archive Record D-LE/D/2/61 John Claxton, 1775.

[38] *Country Life,* 1906.

[39] BMJ 1895.

Auxiliaries to Healing

[40] Hirschfeld, 1785.

[41] Diana Kwon, 'How the Brain Controls Sickness and Health', in *Nature* Vol 614, 2023.

[42] https://www.ecotherapyheals.com/types-of-therapeutic-gardens/ [Accessed 23/02/2023].

[43] George Colman, *The spleen, or, Islington Spa; a comick piece, of two acts, 1776.*

[44] Kentish Gazette, 07 June 1777.

[45] Savidge, 1995.

[46] Tunbridge Wells Standard, 24 June 1881.

[47] Egan, 1819.

[48] An advertisement 'Advice to Bathers at Bath, Hot-Wells, Sea-bathers, &c. &c.' Kentish Gazette, 10 October 1800.
[49] Bisset, 1816.
[50] Mackenzie, 1810.
[51] Pound, 2023.
[52] https://www.greatspatownsofeurope.eu/the-great-spa-towns/anatomy-of-a-great-spa [Accessed 23/02/2023].
[53] Hirschfeld, 1785, translation by the author.
[54] Rowzee, L. 1632: *The Queenes Welles. A Treatise of the nature and vertues of Tunbridge Water* quoted in Gomez Lara, 1993.

Elements of Designed Spa Landscapes
[55] Schulze, K., 'Kurgärten und Kurparks', in Eidloth, 2020. Translation from the German by the author.
[56] Hirschfeld, 1785. Translation the author.
[57] Doyle, 2009.
[58] Kentish Gazette, 07 June 1777.
[59] Stamford Mercury, April 1822
[60] John Betjeman 'English Cities and Small Towns' in Turner, 1945.
[61] *Coventry Herald,* 30 October 1926.
[62] http://bathinbloom.org/bath-in-bloom/baths-achievements (Accessed 20/07/2023). The other finalists in that inaugural year were Aberdeen, Falmouth, Grange-over-Sands, Hove, Middlesbrough, Rhyl and York.
[63] Bristol Evening Post, 04 November 1964.
[64] Hakewill, Plan of Dorton Baths, 1833, Bucks Archives D 215/6/1.
[65] https://londongardenstrust.org/conservation/inventory/site-record/?ID=CRO040 [Accessed 08/08/2021].
[66] https://www.culture24.org.uk/am22917 [Accessed 05/06/2021].
[67] Fiennes, *The Northern Journey and the Tour of Kent (1697)* in Morris, 1982.
[68] Robertson and Hawkins, 1886.
[69] Searle, 1977.
[70] Fiennes in Morris, 1982.
[71] Metellus, *His Dialogue, containing a Relation of a Journey to Tunbridge Wells (1693)* quoted in Gardiner, 1934.

---

[72] Quote of Anthony Hamilton in Jerrold, 1907.
[73] Rabbits, 2018.
[74] Granville, 1840.
[75] Ministry of Housing, Communities & Local Government press release, 20 March 2021, https://www.gov.uk/government/news/ new-raft-of-measures-to-prepare-our-high-streets-and-seaside-resorts-for-summer [Accessed 05/08/2023].
[76] https://cadwpublic-api.azurewebsites.net/reports/listedbuilding/ FullReport?lang=&id=9058 [Accessed 05/09/2021].
[77] http://portal.historicenvironment.scot/designation/GDL00370 [Accessed 03/09/2021].
[78] Shields, 2022.
[79] Cole, 1837.
[80] Hawins, 1976.

SOME BRITISH SPA LANDSCAPES

[81] G.A. Cooke, *A Topographical and statistical description of the County of Kent, etc.* (London: Sherwood, Neely & Jones, c. 1820).
[82] Johnston, 2015.
[83] Metellus, *His Dialogue* quoted in Gardiner, 1934.
[84] Epsom and Ewell History Explorer https://eehe.org.uk/?p=31208 [Accessed 28/08/2021].
[85] Oxford Journal, 26 July 1788.
[86] Granville, 1840.
[87] Heath, 1799.
[88] From a Gardens Trust Zoom talk *Buxton's Pavilion Gardens* by Anna Rhodes of Buxton, Museum and Art Gallery, 18 Oct 2022.
[89] Pittville Park Green Flag Award and Green Heritage Site Management Plan 2016- 2026.
[90] Kent Compendium of Historic Parks and Gardens for Tunbridge Wells Borough: *The Grove* (2009).
[91] South Eastern Gazette, 28 February 1832.
[92] Heath, 1799.
[93] Granville, 1940, describing an old print from early to mid C18.
[94] Egan, 1819.
[95] Leamington Spa Courier, 29 May 1830.
[96] Granville, 1840.

[97] Cunningham, 2007.
[98] Smith & Waller, 1968.
[99] http://friendsofpittville.org.uk/about-pittville/history-of-pittville [Accessed 10/8/2021].
[100] Pound, 2023.
[101] Hanham, 1857.
[102] Hilary A. Taylor, 'Urban Public Parks, 1840-1900: Design and Meaning', *Garden History*, Vol. 23, No. 2 (Winter, 1995).
[103] Joan Warwick, *Down at Beulah* (Norwood Society, 1983).
[104] Watkins, 2002.
[105] Knight, 1833.
[106] Buckinghamshire Gardens Trust Research & Recording Project, *Dorton Spa, Aylesbury Vale DC,* October 2019.
[107] *Act to grant further powers of leasing part of the estates devised and purchased under the will of Sir John Aubrey, baronet [of Dorton House*], 1834, Buckinghamshire Archive Records: D-X_805/5/5.
[108] Dorton Spa, Aylesbury Vale DC. *Buckinghamshire Gardens Trust Research and Recording Project,* 2019.
[109] Hereford Times, 30 June 1860.
[110] Lockett, 2019.
[111] Alison Hodges, 'A Victorian Gardener: Edward Milner (1819-1884)', *Garden History,* Vol. 5, No. 3 (Winter, 1977).
[112] Robertson and Hawkins, 1886.
[113] Samuel Lewis, 'Llandow - Llandyvrydog', in *A Topographical Dictionary of Wales* (London, 1849).
[114] South Shropshire Journal, 08 August 1997.
[115] Scott, 2021.
[116] Doyle, 2009.
[117] https://www.peakdistrictonline.co.uk/matlock-bath-gardens/ [04/01/2025].
[118] Beacon Hotel Cold Bath and Tea Gardens *Kent Compendium of Historic Parks and Gardens for Tunbridge Wells Borough*, 2009.
[119] Curl, 2010.
[120] Ibid.
[121] John Claudius Loudon, *Gardener's Magazine Vol V* (London: Longman, Rees, Orme, Brown and Green, 1829.
[122] Savidge & Bell, 1995.
[123] Hughes, 1995.

Bibliography

24 Views and Scenery of Bath (Bath: Rock and Co., 1877)

British Medical Journal (BMJ)
-Vol. 2, No. 1806 (Aug. 10, 1895) 'Baths And Health Resorts Of The United Kingdom. The Report Of The Royal Medical And Chirurgical Society II'

-Vol. 2, No.1036 (Nov. 20, 1937) 'British Health Resorts Association'

Bisset, James, *Descriptive Guide of Leamington Priors* (Coventry: Merridew and Son, Second edition, 1816).

Borsay, Peter, *A History of Leisure; the British Experience since 1500* (Basingstoke: Palgrave MacMillan, 2006)

Borsay, Peter and Jan Hein Furnée (Editors), *Leisure Cultures In Urban Europe, C.1700-1870: A transnational perspective* (Manchester Scholarship Online: May 2016)

Cole, John, *The History And Antiquities Of Wellingborough: In The Country Of Northampton* (1837)

Conan, Michel (Editor), *Perspectives on Garden History* (Harvard University Press Dumbarton Oaks Research Library and Collection, 1999)

Conlin, Jonathan (Editor), *The Pleasure Garden, from Vauxhall to Coney Island* (Philadelphia: University of Pennsylvania Press, 2013)

Conway, Hazel and Paul Rabbitts, *People's Parks: The Design & Development of Public Parks in Britain* (John Hudson Publishing, 2023)

Cunningham, John (Editor), *An Historical Atlas of Tunbridge Wells* (The Local History Group of the Royal Tunbridge Wells Civic Society, 2007)

Curl, James Stevens, *Spas, Wells & Pleasure Gardens of London* (London: Historical Publications Ltd, 2010)

Denbigh, Kathleen, *Hundred British Spas: Pictorial History* (London: Spa Publications, 1981)

Doyle, Neil, *The Story of St Ann's Well*, (Friends of St Ann's Well Gardens, 2009)

Egan, Pierce, *Walks through Bath* (1819)

Eidloth, Volkmar, Petra Martin, Katrin Schulze (Editors), *Zwischen Heilung und Zerstreuung. Kurgärten und Kurparks in Europa*, (Eschbach: Jan Thorbecke Verlag Gmbh & Co, 2020), [ICOMOS Journals of the German National Committee LXXV]

Elliott, Brent, *Victorian Gardens*, (London: B.T. Batsford, 1986)

Elliot, John, *An Account of the Nature and Medicinal Virtues of the Principal Mineral Waters of Great Britain and Ireland* (London: J. Johnson, 1789)

Ferré, Josep Sánchez, 'Parks and gardens in spa towns. Therapeutic landscape interpretation notes', in *International Symposium on Thermalism and Quality of Life*, (Campus da Auga, Ourense, Spain, 2019)

Fiennes, Celia, *Through England on a Side Saddle in the Time of William and Mary* (London: Field and Tuer, The Leadenhall Press, 1888)

Friends of Valley Gardens, *Valley Gardens Conservation and Management Plan* (2010)

Gardiner, Dorothy, *Companion into Kent* (London: Methuen & Co Ltd., 1934)

Gomez Lara, Manuel J., 'Trotting to the Waters: Seventeenth century spas as cultural landscapes, *SEDERI cultural (*Exwood and Lehman 1993)

Granville, *The Spas of England and Principal Sea-bathing Places* (London: Henry Colburn, 1841) (3 Volumes)

Hanham, Frederick, *A Manual for The Park: or A Botanical Arrangement And Description Of The Trees And Shrubs In The Royal Victoria Park, Bath* (London: Longman, Brown, Green, Longmans and Roberts, 1857)

Hawins, Peter Neville, *The Spas of England* (London: Robert Hale & Company, 1976)

Heath, George, *The New Bristol Guide: ...1066 to 1799...Accounts of the Hotwells and Clifton...* (Bristol: 1799)

Hembry, Phyllis, *The English Spa, 1560 – 1815: A Social History* (London: Athlone Press, 1990)

Hirschfeld, Christian Cay Lorenz, *Theorie der Gartenkunst, Band V* (Leipzig: Weidmann, 1785)

Hobhouse, Penelope, *The Story of Gardening* (London: Dorling Kindersley Limited, 2002)

Hughes, Derek, *The Brine Baths Hotel Nantwich* (Nantwich Museum, 1995)

Jakobsson, Anna, 'Ruled by Routine and Ritual: Spatial Organisation of the Spa Environment at Ronneby, South-East Sweden', *Garden History,* Vol. 32, No. 2 (Winter 2004)

Jerrold, Walter, *Highways and Byways in Kent (London: Macmillan and Co Ltd., 1907)*

Johnston, Mark, *Trees in Towns and Cities. A History of British Urban Arboriculture* (Oxford: Oxbow Books, 2015)

Knight, T., *The History of the Dorton Chalybeate* (Oxford: W. Graham and London: Whittaker and Co, 1833)

Kaspar, Fred, *The Kurgarten: A short overview* (Petersberg: Michael Imhof Verlag GmbH & Co., 2016)

Lamia, 'Modes And Moods: "Dress Does Make a Difference Davy"', *Country Life* Vol. 20, no. 497, Jul 14, 1906

Lasdun, Susan, *The English Park. Royal, Private & Public* (London: Andre Deutsch, 1991)

Layton-Jones, Katy, *National Review of Research Priorities for Urban Parks, Designed Landscapes, and Open Spaces* (English Heritage Research Report Series 4-2014)

Layton-Jones, Katy, *History of Public Park Funding and Management (1820–2010),* (Historic England Research Report Series no. 20/2016)

Lockett, Richard (Editor), *A Survey of Historic Parks and Gardens in Worcestershire* (Hereford and Worcester Gardens Trust, 2019)

Loudon, John Claudius, *An Encyclopaedia of Gardening* (London: Longman, Rees, Orme, Brown and Green, 1826)

Loudon, John Claudius, *Gardener's Magazine Vol V* (London: Longman, Rees, Orme, Brown and Green, 1829)

Mackenzie, George, *A General View of the Agriculture of the Counties of Ross & Cromarty* (London: Richard Phillips, 1810)

Morris, Christopher (Editor), *The Illustrated Journeys of Celia Fiennes 1685 ~ c.1712* (London & Sydney: Macdonald & Co, 1982)

Pound, Christopher, *Legacy of spa gardens,* (Bath: Christopher Pound, 2023)

Plumb, John Harold, *The Commercialisation of Leisure in Eighteenth-century England* (University of Reading, 1973)

Proceedings of the Royal Society of Medicine, -Vol. 13, Balneol. Climatol. Sect. (1920) 'Discussion on the Merits and Defects of British Health Resorts.'

Rabbits, Paul, *Bandstands. Pavilions for Music, Entertainment and Leisure* (Historic England, 2018)

Robertson, William Henry and Miss Hawkins, *A Hand-book of the Peak of Derbyshire, and to the use of the Buxton Mineral Waters* (Buxton: C. F Wardley, 1886)

Roy Strong, *The Renaissance Garden in England* (London: Thames and Hudson, 1979)

Scott, Arthur G., *Strathpeffer 1770 -1900,* (2021)

Savidge, Alan and Rev. Charlie Bell, *Royal Tunbridge Wells,* (Oast Books, 1995)

Searle, M V, *Spas and Watering Places* (Tunbridge Wells: Midas Books, 1977)

Shields, Chris, *The Beulah Spa 1831-1856 A New History* (2022).

Smith, Gordon & Symeon Mark Waller (Ed), *Askern Spa* (1968)

Thacker, Christopher, *The Genius of Gardening. The History of Gardens in Britain and Ireland* (London: Weidenfeld and Nicholson, 1994)

Turner, W. J. (Ed), *The Englishman's Country* (London: Collins, 1945)

Walton, John K. and James Walvin, (eds.), *Leisure in Britain 1780-1939* (Manchester University Press, 1983)

Warwick, Joan, *Down at Beulah* (Norwood Society, 1983)

Watkin, Jeff (Editor), *The Royal Pump Rooms and the growth of Leamington Spa*, (Leamington Spa Royal Pump Rooms Art Gallery and Museum, 2002)

Weisz, George, 'Spas, Mineral Waters, and Hydrological Science in Twentieth-Century France', *Isis*, Vol. 92, No. 3 (Sep. 2001)

Wroth, Warwick & Arthur Wroth, *The London Pleasure Gardens of the Eighteenth Century*, (London & New York: Macmillan, 1896)

Vasset, Sophie, *Murky waters: British spas in eighteenth-century medicine and literature,* (Manchester University Press, 2022)

Index of Places

(Images in **bold**)